A GUID[illegible]

WASHINGTON'S SOUTH CASCADES' VOLCANIC LANDSCAPES

Marge and Ted Mueller

Photos by Bob and Ira Spring

Published by
The Mountaineers
1001 SW Klickitat Way
Seattle, Washington 98134

9 8 7 6 5
5 4 3 2 1

Published simultaneously in Canada by Douglas & McIntyre, Ltd., 1615 Venables Street, Vancouver, B.C. V5L 2H1

Published simultaneously in Great Britain by Cordee, 3a DeMontfort Street, Leicester, England, LE1 7HD

Manufactured in the United States of America

Edited by Dana Lee Fos
Maps by Gray Mouse Graphics
Photographs by Bob and Ira Spring except for the following: pages 21, 42, 45, 50, 70, 90, 128, 150, 153, 196 by Marge and Ted Mueller; page 38 by Hans Carsten, U.S. Forest Service; page 102 by U.S. Forest Service.
Cover design by Watson Graphics
Book design and layout by Gray Mouse Graphics
Typography by Gray Mouse Graphics

Cover photographs: Mount Adams, one of the dominant stratovolcanoes of the South Cascades, as seen from the Dark Divide. *Insets, left to right:* A waterfall pours over a basalt lip in the Muddy River Canyon; Kloochman Rock, near White Pass, is a 1,200-foot-high slab of andesite; a gravel pit at Bunnell Butte shows the structure of a cinder cone.
Back cover photographs, top to bottom: Mount St. Helens and Spirit Lake prior to the 1980 eruption; Mount St. Helens spouts a plume of ash during an early eruption phase; Mount St. Helens and log-covered Spirit Lake two years after the 1980 eruption. (Photos © Bob and Ira Spring)

Library of Congress Cataloging-in-Publication Data

Mueller, Marge.
A guide to Washington's South Cascades' volcanic landscapes / Marge and Ted Mueller ; photos by Bob and Ira Spring.
p. cm.
Includes bibliographical references (p.) and index.
ISBN 0-89886-445-3
1. Hiking—Washington (State)—Guidebooks. 2. Hiking—Cascade Range— Guidebooks. 3. Geology—Washington (State)—History. 4. Volcanism— Washington (State). 5. Washington (State)—Guidebooks. I. Mueller, Ted. II. Title.
GV199.42.W2M84 1995
796.5'1'097975—dc20 95-24080
CIP

Printed on recycled paper

Contents

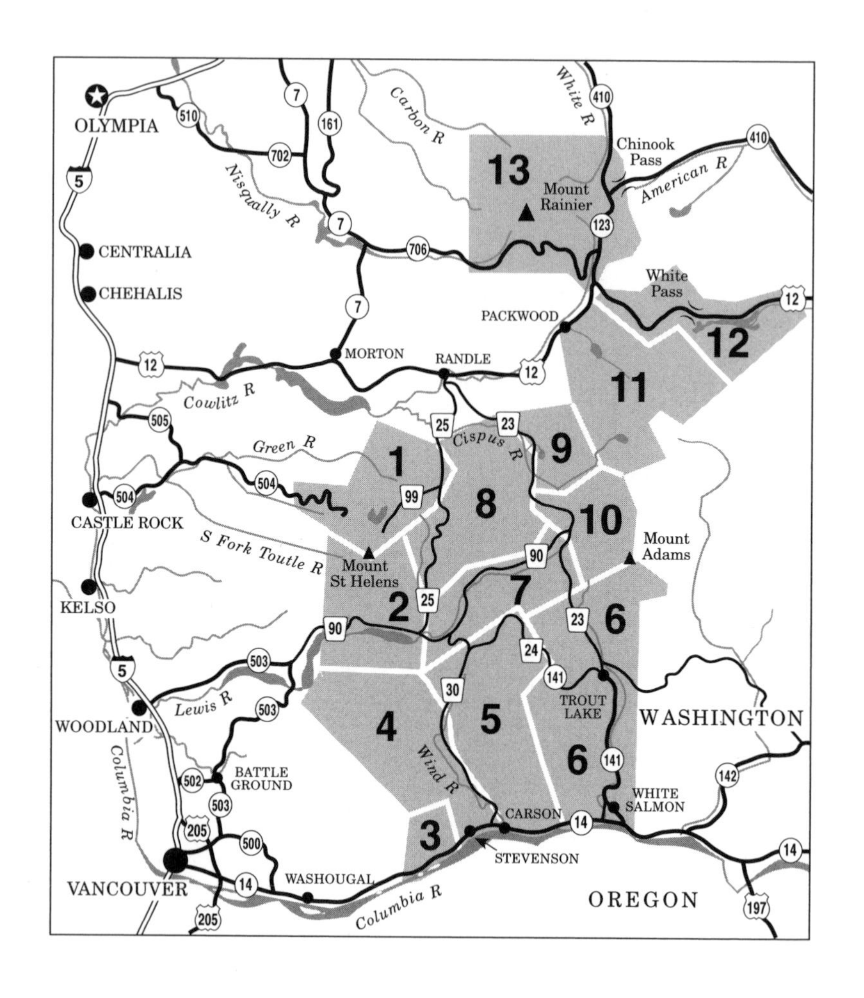
OLYMPIA
CENTRALIA
CHEHALIS
CASTLE ROCK
KELSO
WOODLAND
BATTLE GROUND
VANCOUVER
WASHOUGAL
STEVENSON
CARSON
WHITE SALMON
TROUT LAKE
MORTON
RANDLE
PACKWOOD
WASHINGTON
OREGON
Mount Rainier
Mount St Helens
Mount Adams
Chinook Pass
White Pass
Carbon R
White R
American R
Nisqually R
Cowlitz R
Green R
S Fork Toutle R
Cispus R
Lewis R
Wind R
Columbia R
1
2
3
4
5
6
7
8
9
10
11
12
13

Chapter 5. The Indian Heaven Rift 95

Chapter 6. Mount Adams South 115

Chapter 7. The Lewis River 144

Chapter 8. Dark Divide 150

Preface

While musing over the collective outdoor experiences of more years than our crotchety bodies care to remember, the authors, Marge and Ted Mueller, and the photographer, Ira Spring, concluded that the Gifford Pinchot National Forest of southwest Washington receives far less attention from recreationists than it rightly deserves. Granted, the Mount St. Helens National Volcanic Monument is deluged by more than half-a-million visitors a year; however, few stray beyond visitor centers and roadside interpretive displays. Acknowledged, the scenic value in some sections of forest has suffered from past "timber harvest" (clearcutting)—but the forest still has nearly 175,000 acres of designated wilderness and another 275,000 acres of (as of yet) untouched roadless areas. Admitted, because most of the non-wilderness trails in the forest are open to motorized travel, the noise and wear-and-tear on trails and hiker's nerves by a herd of motorcycles are annoying; however, their raucous presence conveys the impression of greater usage than a head count verifies.

Aside from popular wilderness trails, and weekends and holidays, recreational use of the forest is far lighter than the "take a ticket and stand in line" density at trailheads in such places as the Alpine Lakes Wilderness. Campsites are nearly always available in Forest Service campgrounds, and infrequent road traffic consists mainly of logging trucks and, depending on season, seekers of huckleberries, mushrooms, or wild game. In short, there is a lot of wonderfully uncrowded, delightfully scenic, and geologically fascinating country in the South Cascades of Washington.

It doesn't take much travel in this area to realize one unmistakable fact—This is volcano country! The massive destruction resulting from the 1980 eruption of Mount St. Helens is certainly the most dramatic recent example, but the nearby glacier-clad giants—Mounts Rainier, Adams, and Hood—were also obviously sired by fiery rock from the earth. Many of the lesser mountains in the area, although heavily disguised by forest cover, also have distinctive underlying cone or shield shapes, and sometimes summit craters, that divulge their volcanic origins.

With this observation, we arrived at the criteria for choosing trips for this book and a focus for describing them. Instead of selecting trips solely on the basis of their scenic beauty or recreational value, we chose them because, in addition to being scenic, they led to some interesting examples of the area's intriguing geology. In addition to extolling the beauty of a spot and telling the reader how to get to it, we also tried to explain how that scenery came to be. Ira, the photographer of our team, did demand that the object of the trip also be photogenic!

Because the multi-syllabic precision of a professional geologist is incomprehensible to the uninitiated, we agreed that the text would be, as the beer ads say, "geology lite." It would be written in non-technical terms, in an interesting manner, non-threatening, and hopefully non-boring to the average reader. While the words may lack punctilious scientific precision, within the limits of the vocabulary used, they still accurately describe the geological origin of the region and its interesting features.

LEGEND

U.S. highway	12
state highway	52
USFS primary road	52
USFS connector road	1234
USFS spur road	123
paved road	
improved gravel road	
improved dirt road	
unimproved road	
city, town	
USFS ranger station	
campground	
picnic area	
fire lookout tower	
trail	
cross-country route	
peak	
glacier	
marsh	
lava bed	
unnamed volcanic vent	
named volcanic vent	
crater	
dike	

Acknowledgments

Researching material, taking photographs, and writing a guide book always present a major challenge in physically visiting the areas described. Forest and Park Service changes in plans for, and maintenance of, forest roads and trails and the unpredictable modifications caused by weather, erosion, slides, and other forces of nature can make obsolete the best research over the gestation period of a book.

An additional complication with writing this book was in uncovering background material on the geological origins of features of particular interest because of its special focus on the Cascades of southern Washington. Surprisingly, the geological field research in the area, with the specific exception of Mount St. Helens, is sparse. Aside from a few postgraduate theses dating from the 1970s, the first general geological mapping was completed in the mid- to late-1980s, and detailed mapping and field research in some areas have been undertaken only within the last decade. Much of this material is unpublished, in press, or buried in obscure technical reports. As geologists uncover new information about the area, refine time-dating measurements, and propose theories to explain newly discovered features, opinions change regarding the origins of terrain features and the time periods in which they probably occurred.

The authors always strive for accuracy in their books. We wish to acknowledge persons who have assisted us in achieving this. Especially helpful in providing trail and road information were Jim Nieland and Walt Doan of the Mount St. Helens National Volcanic Monument; Jim Slagle, a Region 6 Trail Engineer; Cliff Bennett of the Mount Adams Ranger District; Dave Olsen of the Randle Ranger District; and Bob Eggett of the Southwest Region of the state Department of Natural Resources (DNR). Valuable background material on the history of the Tract "D" Recreation Area was provided by Steve Andringa of the Yakama Nation.

Jim Nieland also assisted with specific questions regarding the geology of the region, as did Shawn Jones of the Randle Ranger District and Jim Chamberlain of the Gifford Pinchot National Forest Headquarters. Paul E. Hammond, Emeritus Professor of Geology at Portland State University, was an invaluable resource in clarifying our questions on the region's geological features.

A review of draft text of the book for accuracy, always a thankless task, has been provided by the staffs of the Naches, Packwood, Randle, Mount St. Helens, Wind River, and Mount Adams ranger districts. Special thanks are due to John H. Whitmer, Newsletter Editor of the Northwest Geological Society, for his review of all geological material for accuracy.

Spectacular waterfalls, such as Clear Creek Falls near White Pass, are found where streams spill over erosion-resistant layers of basalt.

Introduction

Volcanic fires, glacial ice, catastrophic floods, and the crunching-together of massive tectonic plates combined to form the dramatic landscape that is Washington state today. In the North Cascades the folding is so extreme, and the scraping of immense continental ice sheets so sweeping, that the geological story is a complex puzzle, especially to the casual observer. In the South Cascades nature's handiwork has not been subjected to comparable extreme folding and continental glacier erosion, and the origin of geological features is more easily sorted out. Here the primary (and most readily recognizable) shaper of the landscape has been volcanism, modified by alpine glaciation. Fewer of the most ancient features are exposed—most are covered by thick accumulations of comparatively young volcanic rock, some extruded from volcanoes active within the past 8,000 years. The landscape is no less beautiful in the South Cascades, however, and the geological story is equally fascinating.

Many remarkable geological features can be seen from backroads or even from main highways in the region. The stratovolcanoes of Mount Rainier, Mount Adams, and Mount St. Helens are landmarks from nearly every road. Spiral Butte, an obvious volcano vent, edges Highway 12 at White Pass, while the dramatic columnar formations of the Palisades can be viewed from a highway pulloff west of the pass. On the south side of the area, Mount Mitchell's jagged summit, an eroded core of intruded rock, punctuates the skyline from FR 90 at Swift Reservoir. Beacon Rock is a massive core of a volcano vent that causes people driving on Highway 14, along the Columbia River, to stop their cars and gape in awe.

St. Helens has, of course, garnered the most attention since its violent eruption in 1980. But more chapters in the story of the South Cascades lie along easy paths, backcountry trails, or on beyond in remote areas reached only by cross-country routes. Geological history is recorded in the fascinating lava tubes of the Ape Caves, the stunning beauty of Langfield and Little Goose falls, and the immense, jumbled landscape of Big Lava Bed. Hikers marvel at the dramatic sweep of deeply eroded valleys, the intricate patterns of stratified cliffs, the mystery of sunken craters, and countless other manifestations of Mother Nature's efforts and puzzle how these came to be.

This book takes you to some of the most intriguing and most representative of South Cascades geological phenomena, explaining along the way a bit of how and when they were created. In the following brief overview we fast-forward you through 145 million years of geological history to bring you up to today's landscape. If you didn't take Geology 101 in

Geological Events that Formed the South Cascades

YEARS AGO	
145 – 40 million	Erosion from the North American Plate forms a lowland plain and extends the edge of the continent.
62 – 48 million	Abnormally large suboceanic eruptions build a large seamount (undersea volcano) off the west coast of the North American continent.
50 – 40 million	Collision of the Farallon Plate and North American Plate. Subduction of the Farallon Plate. Continued plate movement scrapes off the seamount and pushes it upward, creating the Olympics and B.C. Coast Range.
40 – 20 million	Fractures on the ocean floor create the Juan de Fuca Plate. As it subducts beneath the margins of the continent, vents beneath a trough on what was then the west coast of the continent begin to erupt.
17 – 12 million	Massive basalt floods flow across eastern Washington and Oregon, reaching to the Pacific Ocean.
12 million	Continental glaciers retreat.
7 million	Uplifting of the South Cascades begins.
3.5 – 1.5 million	Volcanic activity in the Goat Rocks area. Creation of the Goat Rocks stratovolcano.
1 million – 5,000 years	Initial eruptions of present-day Mount Rainier
940,000	Earliest lava flows from the Mount Adams–King Mountain fissure zone.
800,000 – 300,000	Period of major alpine glaciation.
520,000 – 490,000	Large volcano cone built in vicinity of Mount Adams.
460,000 – 160,000	Growth of present-day Mount Adams.
150,000 – 130,000	Period of major alpine glaciation.
130,000 – 120,000	Surge of volcanism builds the shield volcano of King Mountain.
40,000	Volcanic activity begins the formation of Mount St. Helens.
30,000 – 24,000	Lemei and Lake Wapiki vents erupt lava that flows west to Trout Lake and the White Salmon valley.
25,000 – 15,000	Period of major alpine glaciation in the South Cascades.
20,000 – 12,000	Multiple basalt flows and cinder cones are built on the Mount Adams–King Mountain rift.
13,000	Explosive eruptions of Mount St. Helens. The Spirit Lake pluton is intruded.
8,200	Eruption of lava flow that created Big Lava Bed.
500	Lava flows form the symmetrical summit of St. Helens.
15	Eruption of Mount St. Helens, destroys the top 1,300 feet and causes massive debris avalanches and lahars.

college, or if your memory is a bit dusty, you can stop to read the sidebars, which add up to a cram course on the basics. Additional definitions of geological terms are in the glossary at the back of the book.

A Quick Overview of South Cascades Geology

Dramatic Forces Shape the Continent

Long ago—145 million years by geologists' estimation—the edge of the North American continent was located somewhere east of the present-day Washington–Idaho border. Mud and sand eroded from the continent and settled on an offshore shelf, building a low coastal plain that gradually shifted the edge of the landmass westward. These ancient deposits of hardened sedimentary rock, which are the basement of today's South Cascades, can be seen in the vicinity of White Pass, where they have been uplifted and exposed by erosion.

Skip now through an enormous period of time to look at events that began some 50 million years ago. Well west of what was, at that time, the North American continent, abnormally large suboceanic eruptions built seamounts and a huge volcanic plateau on a plate of ocean floor. The ocean plate (known as the Farallon Plate) carrying the seamounts moved northeasterly. At the same time the North American Plate, bearing the continental mass, moved westward. Picture two humongous vehicles—a truck carrying a load of dirt and a bulldozer—hurtling toward each other at the rate of a few inches a year. They meet in a mammoth slow-motion wreck, lasting over millions of years, and become jammed together, with the seamounts crushed against the south end of Vancouver Island, where the landmass had been built up by sediments. Although the edge of the oceanic plate was forced (or subducted) under the bulldozer-like edge of the continental plate, the volcanic plateau was too large to be pushed under, and this top layer was scraped off the ocean plate and pushed over the top edge of the sedimentary rocks. Wedged against the continent, the

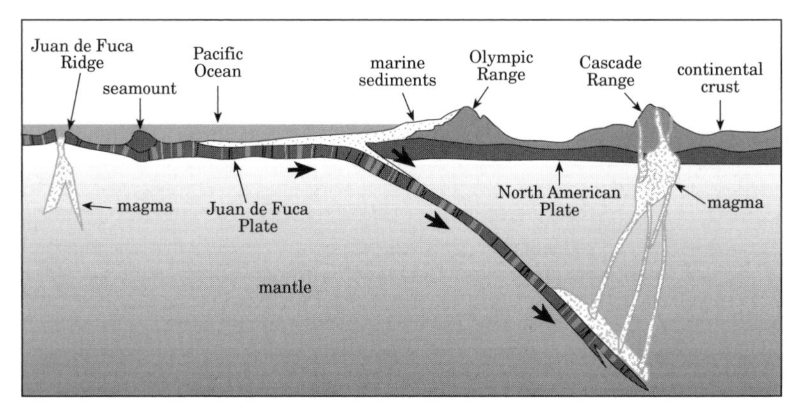

Subduction

volcanic layer accumulated erosion deposits from the land, and over a period of 10 million years both volcanic layers and new sedimentary layers buckled upward by the crushing pressure of the continued movement of the plate. Today these folded rocks form portions of Washington's Olympic Mountains and British Columbia's Coast Range.

Types of Rocks

Rocks are divided into three categories: *sedimentary,* which are formed by deposition of silts, clays, sands, pebbles, or shell fragments by streams, landslides, or underwater processes; *metamorphic,* which form when the crystalline structure of preexisting rocks is altered by heat and/or pressure; and *igneous,* which originate beneath the earth's surface from accumulations of molten magma. Sedimentary and metamorphic rocks are common in the South Cascades and are important in its geological history; however, in this book we use the specific names of only igneous rocks because we are discussing geological features of volcanic origin.

Igneous rocks are divided into two classifications, intrusive and extrusive (volcanic). Unless emplaced near the surface, *intrusive* rocks cool slowly and harden beneath the earth's surface, creating well-structured, grainy rocks with large crystals. In contrast, magma that forms *extrusive* rocks breaks through the earth's surface and cools rapidly, resulting in a much finer crystalline structure, with layering and cooling fractures. Rocks within each of these two classifications are further divided by chemical composition—primarily the percentage of silica they contain. As the silica content increases, the rocks change from black, to gray, to white. The viscosity of molten forms of extrusive rock is also related to silica content; the lower the percentage of silica, the more fluid the lava. The degree of fluidity also has a bearing on the relative violence of volcanic eruptions: the more fluid the lava, the less explosive its eruptions tend to be.

The following intrusive igneous rocks and their extrusive counterparts are referred to in this book. They are ranked, from bottom to top, in the order of increasing percentage of silica:

Increasing Silica	Intrusive	Extrusive
↑	Granite	Rhyolite
↑	Quartz monozonite/granodiorite	Rhyodacite
↑	Quartz diorite	Dacite
↑	Diorite	Andesite
↑	Diabase	Basalt

With this edge of the oceanic plate jammed solidly against the continent, the ocean floor fractured further, creating a new plate (known as the Juan de Fuca Plate). This newly subducting oceanic crust plunged downward into the earth's hot mantle. As it dove, it melted, creating magma pools along an arc beneath the present Cascade crest. Fueled by this fiery source, vents beneath a trough on the coast began to erupt thick layers of lava, fragments of exploded andesite, and ash flows. These eruptions, which continued for some 20 million years, built up layers of lava and volcanic debris more than 22,000 feet deep. This volcanic material accumulated faster than the layers beneath it subsided, and a broad, flat plain of debris and mud flows was created, ringing the vents. Toward the end of this period of eruption, several large bodies of magma from the earth's core intruded into the accumulated native rock to form numerous plutons, dikes, and sills.

The Mountains Begin to Take Shape

This period of violent volcanic events drew to a close about 20 to 18 million years ago when surface volcanic activity slowed. Over the next 10 million years fewer eruptions occurred in the South Cascades; however, even after surface volcanism ceased, intrusions of many small plutons continued, forcing the surface of the land upward. Erosion countered this uplifting, wearing away the surface rock and cutting deep channels into it.

During this time the North American Plate continued its westward drifting. It floated over an unusually large plume of hot material that had bubbled upward from near the earth's core, forming a vast magma pool (hot spot) just below the earth's crust in southeastern Oregon. About 17 million years ago the crust above this magma pool fractured along a fault line near the Washington–Oregon–Idaho border, and in a 5-million-year-long series of events beyond imagination, the hot basalt lava, spreading at a rate of 15 to 25 miles per hour, flooded west across Washington and Oregon. The lava remained fluid long enough to reach the ocean from Hoquiam, Washington, to Newport, Oregon, and covered much of the Cascades along the Washington–Oregon border. This basalt, which was near sea level when deposited, is now found as high as 5,300 feet in the southeastern Cascades—an indication of the magnitude the uplift of the Cascades, which was caused mainly by later pluton intrusions.

Enormous heat, generated as the Juan de Fuca Plate dove under the North American Plate, created a reservoir of magma along an arc beneath the present-day Cascades. About 3.5 million years ago this magma exploited fractures in the surface rock, and over a period of 2 million years a huge stratovolcano, similar to Mount Rainier, was formed south of White Pass. Where is it now? Erosion reduced this one-time giant to its naked inner core—the present-day peaks of the Goat Rocks crest.

Mount Rainier and the other stratovolcanoes now seen in the South

What kind of volcano is that?

The word "volcano" evokes a picture of the glaciated peaks along the crest of Washington's Cascades: Mounts Adams, St. Helens, Rainier, and Baker and Glacier Peak. Actually, these giants are but one type of volcanic landform, called *stratovolcanoes,* or *composite cones*. These are generally large cone-shaped formations built up of many individual layers, each of which may be a lava flow (typically andesite or dacite, which are more viscous than basalt and thus don't spread away from the vent as easily), a collection of volcanic debris, fall-out of volcanic ash and tephra, or a mud flow (lahar). Because stratovolcanoes build up over an extended period of time, they often develop multiple summit vents or "parasite" volcanoes, around their base from branches off their source pool of magma.

Shield volcanoes are far more prevalent in the South Cascades and generally are less dramatic in form. These volcanoes, which are nearly always basalt in composition, develop when the magma source ejects a very fluid form of lava. These flows generally originate from lava fountains that jet as much as 3,000 feet in the air. Droplets from the fountains fall as spatter, coalescing into a lava flow that quickly spreads from the source vent, often as far as many miles, before it cools enough to harden. Some spatter welds together near the vent to make a cone. Because there is only a gradual buildup of lava in the immediate vicinity of the vent, the resulting volcano takes on a shape similar to that of a warrior's shield that has been laid flat on the ground, hence the name. The relatively low profile of these volcanoes makes them harder for the casual observer to identify as a volcanic form, especially when they are heavily forested.

Cinder cones, another classification of volcanoes, result from the accumulation of lava fountain droplets that freeze while airborne and accumulate around the volcano vent. Most of these fragments pile up in the near vicinity, creating steep-walled, symmetrical, cone-shaped formations, seldom more than 1,000 feet higher than the surrounding landscape. Many cinder cones have bowl-like craters in their center, resulting from the subsiding of a lava lake once eruptions have ceased. Large lava flows may often be traced to origins at the base of cinder cones.

Cascades are Johnny-come-latelys on the scene. Rainier grew during a surge of volcanism that began about 1 million years ago. A few hundred thousand years later similar eruptions marked the beginning of Mount Adams, and new vents on major northwest–southeast fault lines in the Mount Adams and Indian Heaven areas extruded large flows of very viscous

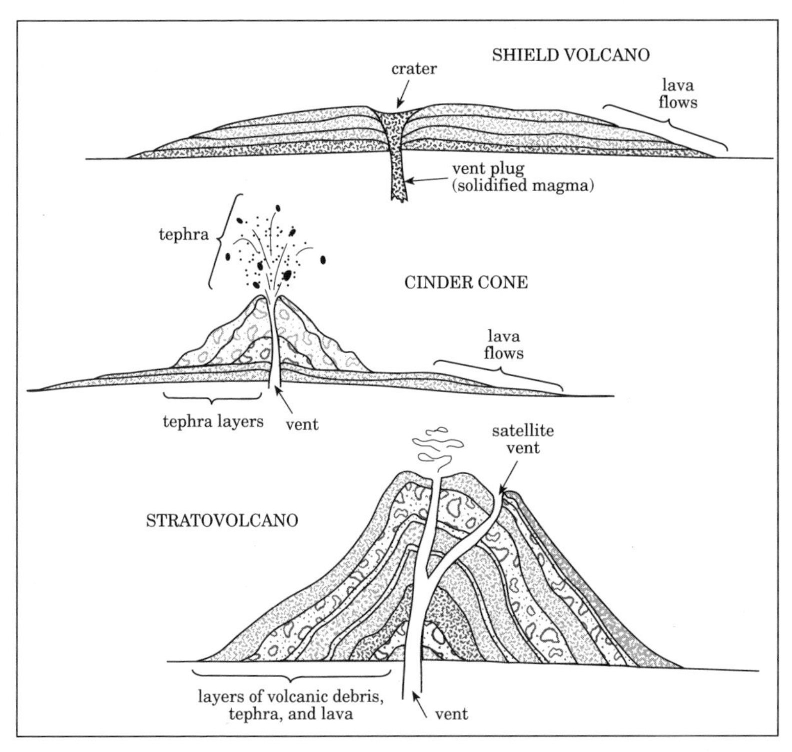

Types of volcanoes

basalt, building up a series of shield volcanoes in those areas.

Except at Mount Adams, Mount Rainier, and the vicinity of the old Goat Rocks volcano, there was then a slackening of volcanic activity and a general uplifting, folding, and faulting of the region lasting until somewhere around 130,000 years ago. At that time another round of volcanism built more shield volcanoes, many of which were later topped by cinder cones. Evidence of this volcanic activity is seen today in the Indian Heaven region with the shield volcano of Bird Mountain, the cinder cones of Red Mountain and Berry Mountain, and north of Mount Adams in the cinder cone of Potato Hill. A major shield volcano developed southeast of Mount Adams at King Mountain about 106,000 years ago.

The youngest of the stratovolcanoes, St. Helens, began its career roughly 40,000 years ago and, like the rebellious teenager that it is, it has erupted at frequent intervals ever since. The last general round of volcanic activity in the South Cascades began about 25,000 years ago with smaller basalt and cinder cone eruptions from a hundred or more vents scattered along the Mount Adams–King Mountain, Mount St. Helens, and Indian Heaven fissure zones.

Plutons, Dikes, and Sills

Geological formations known as plutons, dikes, and sills are all magma intrusions. A *pluton* is a massive, generally dome-shaped intrusion of magma into native rocks. In contrast to other lava flows or volcanic eruptions, the magma rarely breaks through overlying rocks. Instead, it cools slowly beneath the surface, coalescing into coarse-grained rock such as granite, granodiorite, diorite, or gabbro, composed of large crystals. Some compare it to a solidified magma chamber that might once have been the source for eruptive volcanoes. The intrusion of the pluton typically creates extreme pressures and temperatures along its sides, forcing mineral-rich solutions into fissures in surrounding rocks, where they cool to form crystals or veins of minerals. The edges of such an intrusion are, therefore, a likely area for prospecting and mining. Erosion of softer surface rocks eventually exposes this hardened, intrusive core.

Dikes are thin intrusions of magma, like fingers rising off an underlying magma pool, that exploit fractures and weaknesses to cross through bedding planes (layers) in overlying rock. In many cases dikes will solidify into hard, coarse rock ribs within the softer native rock. Frequently dikes break through the overhead rock and act as conduits for lava flows across its surface. Because dikes form in fractures in overlying strata, their presence often indicates areas of stress or weakness in these formations that rising magma may exploit to feed volcanic eruptions. Radial swarms of dikes are found surrounding old, obscure, eroded volcanoes; the common focal point of the dike swarm is generally the central vent of the volcano.

Sills are subsurface intrusions of magma that spread between layers of the overlying rock, exploiting discontinuities or weaknesses between bedding layers. Dikes and sills often occur in tandem; magma presses upward to form dikes and spreads horizontally from those dikes to form sills.

The magma beneath the South Cascades has made its presence known within the past 9,000 years (the wink of an eye, on a geological time scale), when blocky basalt lavas erupted from vents around the base of Adams, in Big Lava Bed south of the Indian Heaven, and at West Crater and Bare Mountain in Wind River country. The most recent event in the South Cascades volcanic history was, of course, the violent explosive eruption of St. Helens in 1980. If the past million years are any predictor of the future, these mountains undoubtedly have not displayed the last of the fireworks issuing from the hot magma pool lying beneath them.

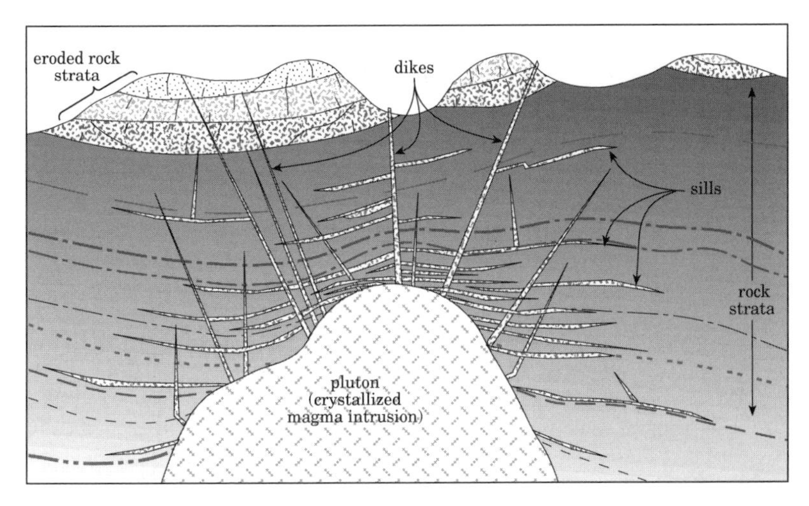

Plutons, dikes, and sills

Glaciation in the South Cascades

Although this book focuses on volcanic origins, glaciers were another major force that dramatically shaped the South Cascades. The vast Cordilleran Ice Sheet, which advanced from Canada and reached its maximum extent about 18,000 years ago, did not extend to the area; its southern limit was in the vicinity of the city of Olympia. However, the climatic cycle that created that ice mass also spawned huge local alpine glaciers that covered much of the South Cascades.

This most recent cycle of alpine glaciation saw several shorter cycles of advance and retreat of alpine glaciers in the South Cascades. Three other major periods of such glaciation occurred in the region. The age of the next oldest cycle is controversial—most geologists list it as between 150,000 and 130,000 years ago. Some suggest this glaciation could have occurred as recently as 60,000 years ago; still others feel 315,000 to 300,000 years would be a better date range. The third period of glaciation is currently placed at somewhere between 800,000 and 300,000 years ago. The oldest identified glacial cycle is presently dated at about 1 million years old.

The dates of glaciation are important because one means of geologically dating a volcanic feature depends on whether or not the surface shows erosion from a given period of glaciation, which indicates it existed prior to that period. Because the accepted dates of glacial periods still vary widely, glaciation is responsible for a good deal of the ambiguity in determining the age of formations. Glaciation further complicates geological dating because it obliterates or rearranges many of the surface features and leaves others covered with thick, obscuring layers of glacial till, making it difficult at times to find exposed underlying rock for examination.

How to Use This Book

This book is not just a trail hiking guide to the South Cascades. Many interesting geological features can be reached or seen from forest roads in the area, so some of the trips included require no hiking at all. Some are short, easy strolls, and a number include barrier-free paths. At the other end of the spectrum, some fascinating spots have not, and probably never will have, a trail built to them. They require cross-country travel, woodsmanship, and wilderness navigation skills to reach them and get back safely.

Information Summaries

Information summaries at the beginning of each hike description provide a brief overview of the road trip, hike, or cross-country route.

Trails are listed at the beginning of hikes. Some Forest Service publications use names, some use numbers. Both the names and numbers of trails are shown, when both exist, so the reader can readily cross-reference these publications.

Rating includes the difficulty of the trail and the usage restrictions. The following codes indicate the difficulty of hikes and cross-country routes:

(E) Easy. Requires few skills and offers little physical challenge. Streams are always bridged.

(M) Moderate. Requires moderate hiking skills and/or offers moderate physical challenge. The tread may contain roots and rocks, and trailside vegetation is generally cleared as a result of use. Changes in elevation are moderate. Streams are most often crossed by bridges or foot logs.

(D) Difficult. Requires a high degree of hiking skills and/or is physically challenging. The tread is seldom graded, and vegetation can be expected to encroach on the trail. Elevation change may be severe. Streams may need to be forded. Routefinding skills may be required.

(X) Exposure. These trails are not necessarily difficult—in fact, some are relatively easy hiking, but the vertical exposure may be uncomfortable for persons with acrophobia.

Note that this ranking does not precisely follow the Forest Service classifications that you will find published in their recreational opportunity guides and posted at some trailheads. Their trail categories are Easy, More Difficult, and Most Difficult, which are roughly parallel to the first three categories used here.

In the Ratings listing, the difficulty of the trail is followed by the usage restrictions. Travel on any specific trail may be open to use by hikers, mountain bicycles, saddle and pack stock, llamas, or motorcycles, or any combination thereof. Motorized travel and bicycles are *never* permitted in Designated Wilderness Areas.

Distance and **Elevation** show the hiking distance from the trailhead

Llamas, which are permitted on a number of trails, are popular pack animals because they have a lower impact on trails than horses.

to the destination and the elevation at the start and finish. Occasionally distances to intermediate points and their elevations are shown. This information is primarily for the use of hikers planning trips, thus it is not shown for road trips or short paths.

Maps

Several types of maps and guides are of value to hikers and road travelers. Each has its pros and cons. The **USGS Quadrangle 7½-minute series topographic maps** (scale 1:24,000) are most accurate with respect to topographic information but can be woefully out-of-date with regard to current forest roads and trails. The plus side of this is that some may show old abandoned trails that are still followable but do not appear on current maps.

Forest Service Ranger District maps (scale 1:63,600) also show topography contours, although their smaller scale may require a magnifying glass to read the contours. These maps generally are reprinted every four to five years, and road and trail information is reasonably current (except that roads or trails the Forest Service would like to forget somehow, conveniently, disappear). The maps show complete detail of all forest roads of any type within the district, keyed to quality and surface. They may be purchased at district ranger stations.

Green Trails 15-minute maps (scale 1:69,500) are commercially published maps that show contour lines at a less-frequent interval than 7½-minute maps. They are kept up-to-date with road and trail information from frequent field checks.

For areas in this book, overall **National Forest Visitor Maps** (scale 1:126,720) of the Gifford Pinchot or Wenatchee National forests, which may be purchased at ranger stations or at many recreational equipment and sporting goods stores, are desirable aids for traveling forest roads. These maps provide a good overall view of the area and show most primary and connector roads, as well as some road spurs.

Trail Guides and Recreational Opportunity Guides printed by each ranger district show sketch maps and have brief descriptions of trails within the district. These are available for free or for a nominal charge at district ranger stations.

For those interested in the geology of the area, the most convenient maps are the state **Department of Natural Resources geological quadrangles** (scale 1:100,000), or a nice color summary of these in the Geologic Map of Washington—Southwest Quadrant (scale 1:250,000). These are all dated 1987 and are available for a reasonable fee from the DNR in Olympia.

Forest Road Travel in the South Cascades

Roughing It in the Backcountry

Most of the destinations described in this book require leaving comfortable, paved highways with service stations and telephones to travel forest roads of varying quality and questionable maintenance, often a couple of hours or more from the nearest support facilities. Most forest roads were initially funded by and built for one purpose: access to logging sales units. They were not engineered for speed or for low-slung passenger cars; drivers may find them uncomfortable and unnerving. Primary or connector roads in the forests, even paved ones, are often single-lane, with pullouts every few thousand yards to permit head-on traffic to pass. Logging trucks and other large commercial vehicles use citizens band radio to advise each other of their locations so that one or the other can be prepared to use a pullout before a head-on confrontation occurs. The CB channels used on any particular stretch of road are generally posted near the start of the road, and passenger vehicles with CBs would be well advised to monitor the appropriate channel, or be prepared to back up to the nearest pullout—you can bet a logging rig isn't about to!

Gas stations and mechanics are non-existent once you are off state or federal highways. Always fill your tank before leaving main highways. Make sure your vehicle is in good mechanical condition and that you are equipped to handle simple emergency situations such as flat tires or low water or oil. Check at ranger stations regarding road conditions and closures before venturing into remote areas.

Road Numbering

Forest roads are identified in this book by the initials FR followed by a two-, four-, or seven-digit number that, most of the time, will also be found on a sign or numbered post where the road starts or branches off another road. The numbering system is hierarchical—that is, FR 5420670 will be a road branching off FR 5420, which in turn will have branched off FR 54. A road numbered FR 5400670 would branch directly off FR 54, with no intermediate road.

In general, the smaller the number, the better quality the road. Primary Forest Service roads, designated by a two-digit number, usually are 1½ lanes or wider, with a better surface (sometimes paved) and reasonably well maintained (although some gravel ones can become quite washboardy between maintenance cycles).

Roads with four-digit numbers are Forest Service connector roads. These are most often a generous 1-lane gravel with room for careful passing and are reasonably well maintained. Roads with a seven-digit number are spur roads that are somewhat of a grab bag (and are generally referred to by the last three digits—for example, "the 670 road"). These can vary from a fairly comfortable single-lane gravel to a narrow, rutted, potholed nightmare, to a bulldozer track passable only by a high-clearance four-wheel-drive vehicle.

Trail Hiking, Cross-country Travel, and Summit Climbs

Most hikes in this book take you to a specific feature or area by the shortest route possible and generally return you the same way. Where there are loop options of about the same distance and difficulty, these are described. With a few exceptions, all the hikes in the book are designed to be done in one day (driving time to the trailhead excluded). This means that most are from a short mile to 8- to 10-mile round trips to and from a feature or view, although a given trail may continue on much farther than that. In some areas, such as the Indian Heaven Wilderness or the Goat Rocks, it is possible to combine several individually described hikes into a trip of two days or more.

Formal trails in the South Cascades vary widely in maintenance and quality. Trail work is dependent upon a limited budget and the availability of trail crew and volunteer labor. As a result, heavily used trails generally get first call for maintenance, while the seldom-visited ones may receive little or no work.

Wilderness permits are required for either trail or off-trail travel in any Wilderness Area. At the time of this writing, these permits are available at

Mountain bikers follow the Ape Canyon Trail along the edge of the Muddy River lahar. The south side of Mount St. Helens is in the distance.

trailheads or ranger district offices; should increasing visitor pressure on wilderness areas lead to limits to the number of parties on a given wilderness trail, permit rules may change.

Trips included in this book involving cross-country travel require a greater level of backwoods skills than trail hiking. Such trips should not be made alone; loss of direction or an off-trail injury without assistance would be a major problem.

In general, off-trail travel is not encouraged in Washington's backcountry, because of the possibility of damaging fragile areas. The off-country routes described in this book traverse lava beds, loose tephra, old clearcuts, and other such terrain that will not be damaged by boot treads. Nonetheless, at all times use care to not destroy plant life or in any other way damage or permanently mark the terrain. The terrain in the restricted areas on the north side of the Mount St. Helens National Volcanic Monument is extremely fragile. A stiff fine is imposed for even stepping off the trail in these areas.

Summit climbs on any of the major volcanoes require prior experience in mountaineering and specialized gear. Permits are required to climb

St. Helens; climbers attempting to summit Adams or Rainier must check in and check out at the ranger stations.

Safe hiking and cross-country travel require good boots and suitable clothing. Hikers should be familiar with the use of map and compass for wilderness navigation and should always carry a backpack or rucksack with the Ten Essentials:

1. Map
2. Compass
3. Flashlight or headlamp with spare bulbs and batteries
4. Extra food
5. Extra clothing
6. Sunglasses
7. First-aid kit
8. Pocket knife
9. Matches in a waterproof container
10. Fire starter

Some of the exposed, higher altitude trails, such as the Pacific Crest Trail in the Goat Rocks, can be potentially hazardous, depending on season and weather, and require even more experience, preparation, and equipment to assure a safe and pleasant trip. Magnetic anomalies caused by local geographical features, such as have been reported in Big Lava Bed, may make compass courses inaccurate, so on long cross-country trips route flagging may be advisable to assure finding your way out on the return trip (remove flags on the way out, to prevent confusion for future parties). Gear such as ice axes, crampons, ropes, carabiners, pitons, wands, and altimeters are prerequisites for summiting any of the major volcanoes.

A Note About Safety

Safety is an important concern in all outdoor activities. No guidebook can alert you to every hazard or anticipate the limitations of every reader. Therefore, the descriptions of roads, trails, routes, and natural features in this book are not representations that a particular place or excursion will be safe for your party. When you follow any of the routes described in this book, you assume responsibility for your own safety. Under normal conditions, such excursions require the usual attention to traffic, road and trail conditions, weather, terrain, the capabilities of your party, and other factors. Keeping informed on current conditions and exercising common sense are the keys to a safe, enjoyable outing.

The Mountaineers

Top: *Mount St. Helens and Spirit Lake prior to the 1980 eruption*. Bottom: *A cloud of steam and ash rises thousands of feet above the mountain during one phase of the eruption.*

1

Mount St. Helens North

A cataclysmic blast registering on seismographs around the world, and spreading volcanic debris into the stratosphere, shattered Mount St. Helens in May of 1980. The strikingly beautiful, symmetrical cone that had been built up over the past 2,000 years was destroyed when the top 1,300 feet of the 9,600-foot-high peak was decimated, leaving a gaping, ragged stump. The north side received the brunt of the destruction. In minutes the entire landscape was altered within a 15-mile-wide arc sweeping north from the peak.

This most recent eruption of the mountain should have come as no surprise; throughout its existence this youngest of the South Cascades stratovolcanoes had a legacy of violent eruptions, interspersed with dormant periods. Scientists knew the possibility of an eruption but could not predict when it would occur, nor its specific nature and intensity.

Through geological studies, the mountain's history can be traced back about 40,000 years when a series of explosive lava flows issued from a northwest–southeast fault zone running through the summit of the present mountain. Since that time the mountain's history has been characterized by periods of explosive eruptions, lava flows, and extrusions of lava domes ranging from 15,000 years to as brief as 500 years, followed by dormancy.

The present-day volcano began to take shape some 2,200 years ago. At that time the composition of lava flows changed abruptly from dacite to andesite and basalt, although new dacite domes also were extruded. The Cave Basalt lava flows, in which current lava caves are found, were extruded during this time. After 600 years of volcanic activity the mountain again fell silent. About 1,200 years ago a brief series of eruptions caused the extrusion of the Sugar Bowl, a dacite dome on the mountain's north flank, and was accompanied by a powerful lateral blast, flows of hot debris, and lahars.

A mere 500 years ago an explosive eruption spewed clouds of pumice northeast across Washington and Canada, and massive lava flows moved down the north, south, and west sides of the mountain. A dacite dome extruded at this time created the symmetrical summit of the peak as it appeared prior to the 1980 eruption. St. Helens briefly slumbered until about 1800 when several explosive eruptions on the north flank were accompanied by the extrusion of a dacite dome on the north side of the mountain, later called the Goat Rocks. Although small eruptions were witnessed in 1831, 1835, 1842, and 1847, St. Helens remained relatively quiet until its devastating blast of 1980.

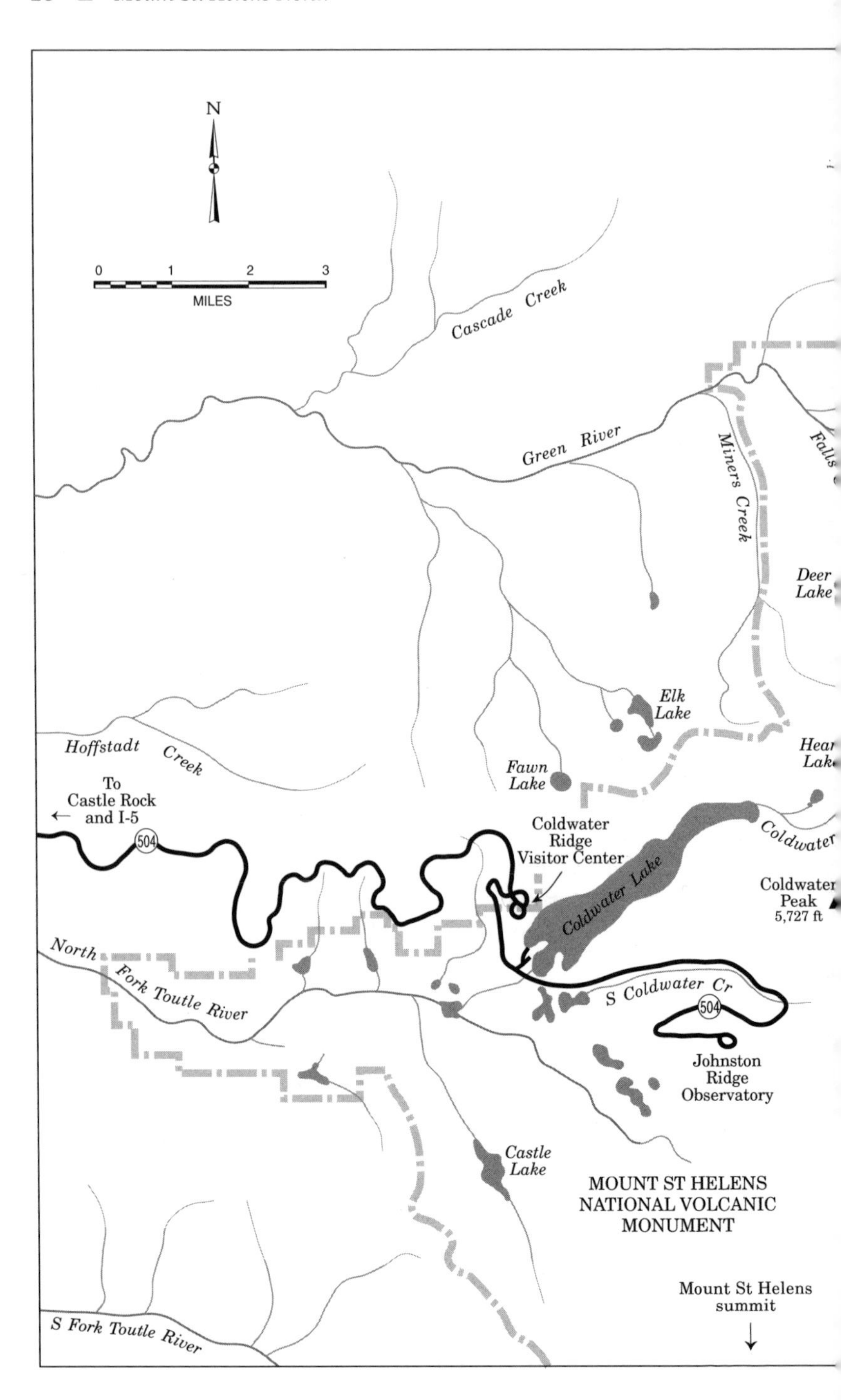
N
0
1
2
3
MILES
Cascade Creek
Green River
Miners Creek
Deer Lake
Elk Lake
Hoffstadt Creek
Fawn Lake
To Castle Rock and I-5
504
Coldwater Ridge Visitor Center
Coldwater Lake
Coldwater Peak 5,727 ft
North Fork Toutle River
S Coldwater Cr
504
Johnston Ridge Observatory
Castle Lake
MOUNT ST HELENS NATIONAL VOLCANIC MONUMENT
Mount St Helens summit
S Fork Toutle River

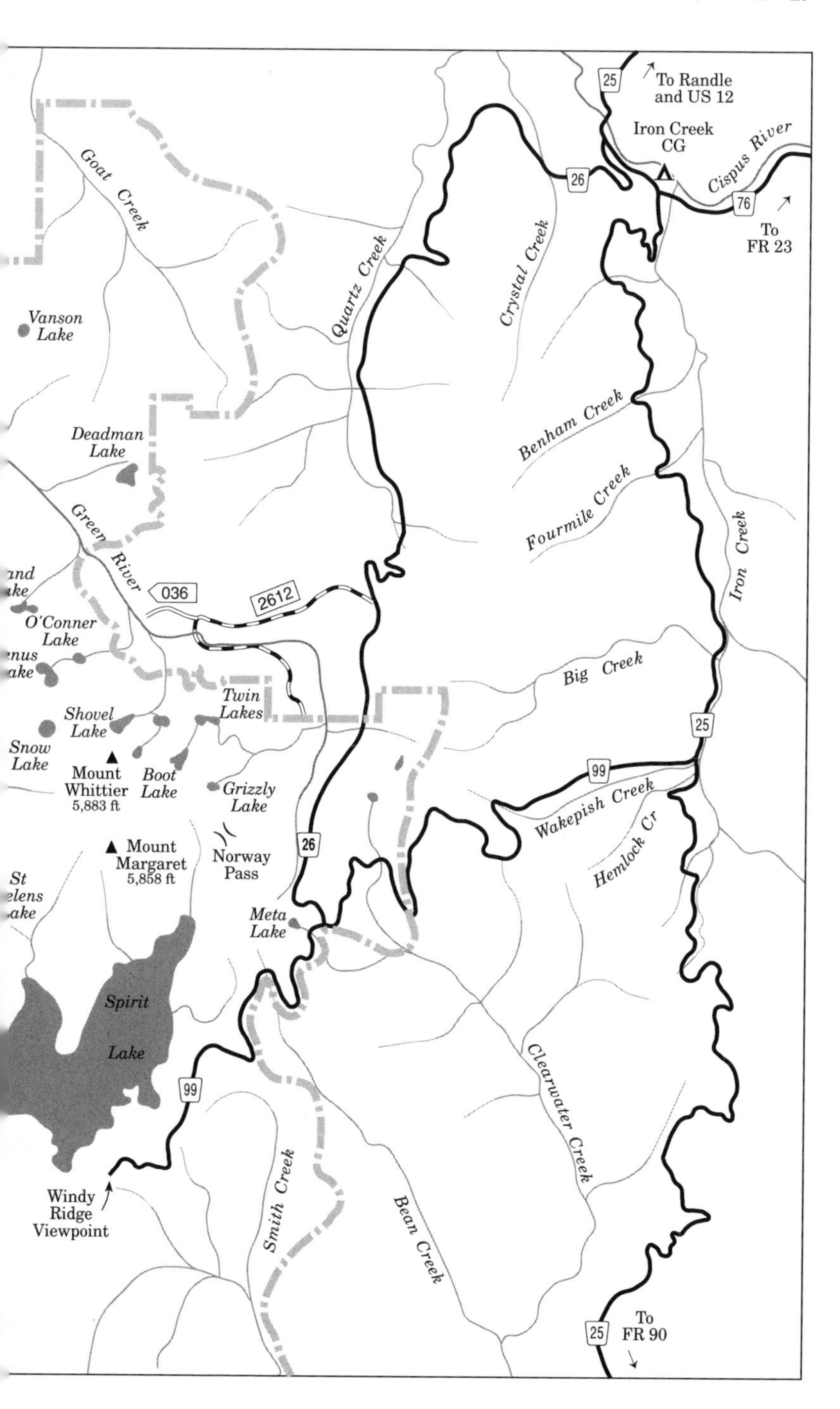

25
To Randle and US 12
Iron Creek CG
Cispus River
26
76
To FR 23
Goat Creek
Quartz Creek
Crystal Creek
Vanson Lake
Benham Creek
Deadman Lake
Fourmile Creek
Green River
Iron Creek
036
2612
O'Conner Lake
Big Creek
Twin Lakes
Shovel Lake
25
Snow Lake
Mount Whittier 5,883 ft
Boot Lake
99
Grizzly Lake
Wakepish Creek
Hemlock Cr
Mount Margaret 5,858 ft
Norway Pass
26
Meta Lake
Spirit Lake
Clearwater Creek
99
Smith Creek
Bean Creek
Windy Ridge Viewpoint
25
To FR 90

Some Major Geological Events in the Mount St. Helens Region

YEARS AGO	
50,000 – 36,000	Series of explosive flows occur from a northwest–southwest fault zone running through the summit of the present Mount St. Helens.
25,000 – 15,000	Period of major alpine glaciation.
20,000 – 18,000	Explosive eruptions, lava flows, and extrusions of lava domes. The Spirit Lake pluton and the granodiorite dome that is now Mount Mitchell are extruded.
13,000 – 10,000	Explosive eruptions and several pyroclastic flows with ejections of pumiceous tephra that falls as far east as central Washington.
3,900 – 3,300	Explosive eruptions and pyroclastic flows. Huge volumes of ash and tephra fall as far north as Canada.
2,900 – 2,500	Eruptions of volcanic debris. Lahars flow north into the Cowlitz River via both forks of the Toutle River and south to the Lewis River.
2,200 – 1,600	Eruptions. Lava composition changes abruptly from dacite to andesite and basalt, although new dacite domes also are extruded. Present-day volcano begins to take shape. The Cave Basalt is extruded.
1,200	Brief series of eruptions causes the extrusion of the Sugar Bowl, a dacite dome on the mountain's north flank. It is accompanied by a powerful lateral blast, flows of hot debris, and lahars.
1,000	Butte Camp Dome forms from a magma intrusion.
500 – 350	Explosive eruptions; massive lava flows move down the north, south, and west sides of the mountain. Formation of the symmetrical summit. Worm Flows created on south side of mountain.
200 – 140	Explosive eruptions on north flank of St. Helens. Extrusion of dacite dome on north side of mountain.
15	Major eruption of May 18, 1980 removes the top 1,300 feet of Mount St. Helens' summit and generates huge debris avalanches and lahars.

On March 25, 1980, a swarm of moderately strong earthquakes shook the mountain over an eight-hour period, triggering avalanches and rockfalls and opening fractures in glaciers high on the peak. Earthquakes continued, and on the 27th an explosion was heard, and a column of ash rose from a newly formed summit crater. More explosions occurred over the next few days, and a second crater opened. Seismic activity continued, and over the next two weeks the craters merged and the mountain developed a bulge on its north side. Eruptions then ceased, and throughout most of April the only activity was continuous venting of fumaroles from the enlarging crater.

On May 7 small eruptions resumed in the crater, spewing ash over the lower slopes and creating steam clouds above the summit. Several new

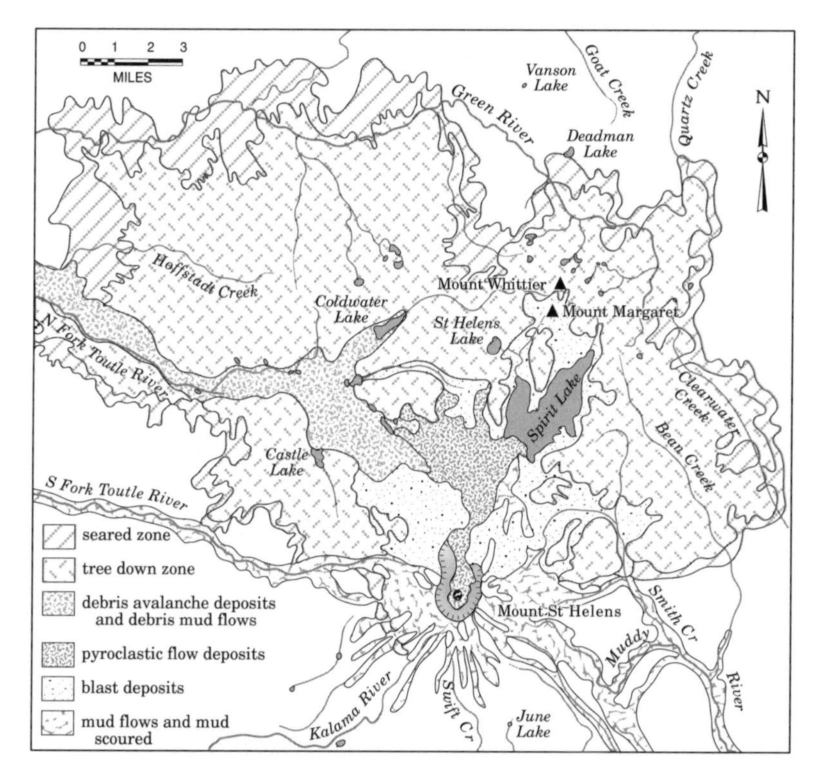

Features of the 1980 eruption of Mount St. Helens

steam vents opened on the upper slopes around the perimeter of the crater, and the bulge on the north side of the mountain became more pronounced, with the flank of the mountain moving outward to the north at a rate of about 5 feet a day. Earthquakes continued, with local tremors averaging twenty to forty daily.

On May 18 at 8:32 A.M. an extremely strong earthquake (magnitude 5 or higher) shook the mountain; the entire north side quivered with the consistency of gelatin, then broke away in three massive blocks, totaling more than ½-cubic-mile in volume—the largest such avalanche in recorded history. This massive debris avalanche swept down into the North Fork of the Toutle River, banking off ridges to the north and sometimes overriding them. The huge wall of earth displaced Spirit Lake to the northeast, raising its bed by more than 180 feet. At a breakneck 60 miles per hour the earth wall raced down the Toutle River drainage for more than 16 miles. In the vicinity of Coldwater Creek, avalanche debris accumulated to a depth of 300 to 450 feet.

With the overburden of the north flank suddenly gone, the core of the volcano exploded with a force equal to 30 million tons of TNT. The lateral blast, aimed to the north, was like pressure relieved from an enormous seltzer bottle. The blast wave, with temperatures as high as 400 degrees Fahrenheit, traveled at more than 650 miles per hour and carried a searing blanket of dirt, gravel, and debris particles with it, destroying everything within 6 miles to the north of the former summit. Forests on that side of the mountain, up to 15 miles away from the crater, were sandblasted and leveled by the force of the discharge, and trees on the perimeter were scorched by its heat.

The blast was followed by pyroclastic flows of hot fragments (between 600 and 1,400 degrees Fahrenheit) from the magma heart of the volcano, which overlaid the debris avalanche for up to 4 miles north of the crater. Six more of these flows occurred during the six months following the May eruption.

The heat of the eruption immediately melted the upper 45 feet of the glaciers that blanketed the peak; the resulting water, mixed with avalanche debris, pumice, and ash turned into lahars. These rivers of warm, viscous mud with the consistency of wet concrete, rushed down the slopes and former drainages of the mountain at a rate of up to 25 miles per hour, scouring to bare earth anything in their path. The flows continued downstream to the Columbia River, where deposited sediments disrupted shipping channels in the river for several months.

Fifty-seven lives were lost in the blast and its aftermath, more than 300 homes were destroyed, countless numbers of animals died, and 96,000 acres of trees were toppled. A column of ash, rising as high as 12 miles above the mountain during the eruption, was carried east by prevailing winds, circling the globe. Areas as far away as western Montana were blanketed with 2 to 5 inches of a powdery deposit of ash, raising havoc for several weeks with transportation, water reservoirs, sewage plants, and even breathing.

The eruption of Mount St. Helens triggered enormous interest worldwide. Scientists were drawn here to study the natural disaster and its effects on the environment. An ever-growing stream of curious visitors arrived to view the scene. In 1982, Congress, recognizing the necessity of preserving this unique environment for scientific research, visitor education, and recreation, created the Mount St. Helens National Volcanic Monument.

Several roads provide access to the monument, and visitor viewpoints and interpretive centers have been established at numerous points on the approaches to the mountain.

There is no direct road route across the north side of the monument, where more than 20,000 acres remain roadless (even though some economic interests have lobbied for a connection). This area includes long ridges joining some of the most rugged peaks in the monument, a host of alpine lakes, three major river and stream drainages, blast-flattened forest, and the only pocket of still-standing forest on the north edge of St. Helens.

To reach the northwest side of the mountain, the major arterial is Highway 504, which runs east from I-5 near Castle Rock, passes the visitor center at Silver Lake at 5.4 miles, continues on to the Coldwater Ridge Visitor Center at 45 miles, and ends at the Johnston Ridge Observatory at 53 miles.

Reach the northeast side of the monument from I-5 via US 12 to Randle, 50.4 miles, then head south on Highway 113, which becomes FR 25 at the Gifford Pinchot National Forest boundary. FR 25 continues south across Elk Pass (closed in winter) to join FR 90 67 miles south of Randle. Highway 503 and FR 90, 47.9 miles from I-5, provide an alternate approach to FR 25 from the south.

Two forest roads lead west from FR 25 into the monument. The most heavily used is FR 99; it leaves FR 25 30 miles south of Randle and continues west and south for 17.4 miles to the road-end viewpoint at Windy Ridge. Several parking spurs for trailheads and interpretive sites are passed along the way. Interpretive programs are presented at Windy Ridge during summer months. A second access road is FR 26; it leaves FR 25 8.5 miles south of Randle and follows a twisting path southward up the Quartz Creek drainage to eventually join with FR 99 9.3 miles west of FR 25.

Excellent interpretive visitor centers are at Silver Lake, 5 miles east of Castle Rock at North Fork Ridge, 34 miles east of Castle Rock and at Coldwater Ridge and Johnston Ridge near the mountain. A network of trails of varying difficulty leads walkers and hikers to eye-popping vistas of the mountain's shattered crest and the surrounding devastation. A climbing permit is required to go to the crater rim. Trails that lead to some of the most interesting geological views are described here.

North Fork of the Toutle River

Hiking trail through a debris avalanche

Trail: Hummocks Trail #229
Rating: (M); hikers
Distance: 2.5 miles
Elevation: Trailhead 2,400, low point 2,260 feet
Maps: USGS Elk Rock, USFS Mount St. Helens Ranger District
Driving Directions: Take Exit 49 from I-5, and head east on Highway 504. The Mount St. Helens Visitor Center, in 5.4 miles, is well worth the stop for its interpretive displays and programs. Continue east on Highway 504 another 39.6 miles to the Coldwater Ridge Visitor Center. Several roadside pullouts en route offer ever-expanding views of St. Helens and other geographical features of the monument. After viewing the interpretive displays and programs at Coldwater, continue south on Highway 504 for 2.1 miles to the start of Trail #229.

Hummocks is an especially appropriate name for this trail, because it threads through piles of the debris avalanche of rock that slid from the north side of St. Helens in its 1980 eruption. For the casual sightseer, a short hike leads to a Y-junction and a short spur to a viewpoint overlooking the hummocky moonscape and the present channel of the North Fork of the Toutle River. This dramatic vista will whet your appetite for the longer loop trail through wild, colorful terrain.

The pumice path of the Hummocks Trail heads southeast and soon reaches the rim above a green pond. This is what remains of a small lake that, following a wet winter and spring a few years ago, drained away into the Toutle River after breaching the soft, sandy dam that had originally created it. The trail drops to the present shoreline of the pond, where marsh plant communities are regenerating and a mob of pollywogs abounds in late spring.

The scenery is reminiscent of a small Dakota Badlands as the trail threads through 50- to 100-foot-high levees and hummocks of wildly varying pastel shades, each color marking a different phase in the mountain's history of eruptions. The light gray stems from recent eruptions within the past 200 years; the pink is older, probably from 2,500-year-old eruptions; the rust-red and white from the eruptive cycle 500 to 300 years ago; the black from volcanic activity 2,200 to 1,700 years ago. Strewn about the surface are blocks of lava, some up to TV set size. At one spot the path passes a pair of blast-peeled tree trunks, 3 feet in diameter, that bear mute testimony to the force of the eruption. One stands upright but is fractured off; the other is grotesquely root-end up, buried upside down in the debris.

The trail winds past a meadow that has regenerated on the once-barren surface. Approach quietly—elk are frequently seen here. It then bends west and drops to pass more ponds. Some scattered hardy pines and cedars have taken root, and willows now crowd the banks of the ponds. A pair of bowl-shaped holes, one about 30 feet deep and 200 feet across, are believed to have been formed by blocks of glacial ice immediately following the eruption; the ice melted to leave the depressions.

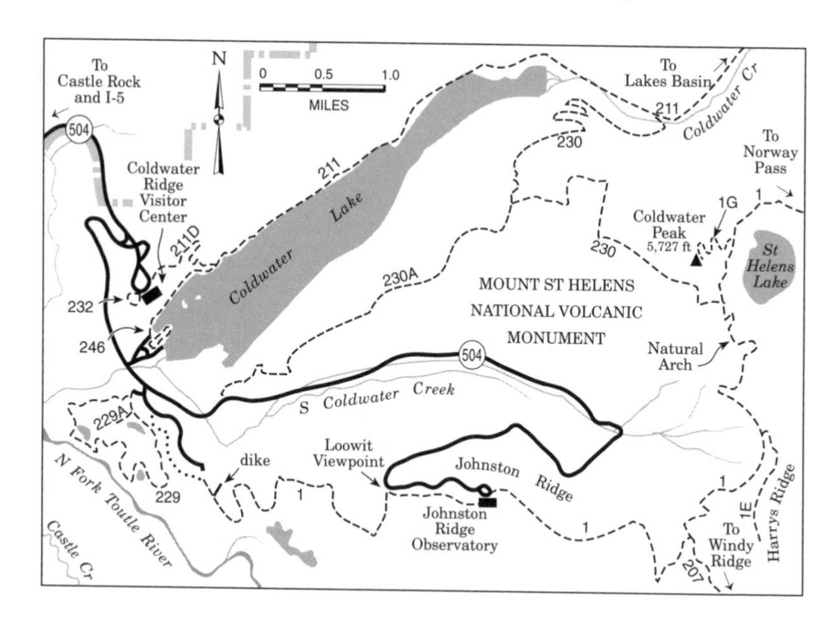

Hummocks Trail winds through a massive debris avalanche.

The route now heads west along the edge of a white debris cliff above the current Toutle River drainage, then descends 100 feet down a stream-cut gully to a flat rocky plain at the current river level. After tracing the base of this debris cliff downstream for a few hundred yards, switchbacks regain the cliff top near a pair of pointed hummocks, one pink, the other pale gold. One more pond is passed before the path climbs uphill once again to join the outbound trail near its scenic viewpoint.

Boundary Trail (West End)

Hiking trail with views of a dike, the Pumice Plain, and the crater of Mount St. Helens

Trail: Boundary Trail #1
Rating: (M, X); hikers
Distance: 3.2 miles
Elevation: Trailhead 2,510 feet; Loowit Viewpoint, west of Johnston Ridge Observatory 3,920 feet
Maps: USGS Elk Rock, Spirit Lake West; USFS Mount St. Helens Ranger District
Driving Directions: Follow directions to the North Fork of the Toutle River, preceding. The western trailhead for Trail #1 is at the end of a spur road 0.2 mile beyond the Trail #229 parking lot. The east end of this trail segment, Loowit Viewpoint, is 0.8 mile west of the Johnston Ridge Observatory. (As of 1995 the observatory, the road to it, and this section of the Boundary Trail are under construction. They are expected to be completed in 1996.)

When the 1980 eruption devastated the landscape north of the mountain, it obliterated all of the existing trails in the area. As these trails were restored and new ones were created, trail planners took advantage of the blank slate left by the eruption and routed paths past spots with prime views of the crater and interesting ecological features exposed by the blast. Boundary Trail #1 offers some of the best overall views of the results of the eruption.

Boundary Trail is a 63-mile-long National Recreation Trail that runs east to west between Mount St. Helens and Mount Adams. The route follows the old boundary between Rainier National Forest and Columbia National Forest, thus its name. These two Department of Agriculture units were created in 1908 to manage federal lands in southwest Washington; however, since 1949 this area is a part of, or is managed by, the Gifford Pinchot National Forest.

The trail section described here is at the west end of Boundary Trail #1. Because of construction the trail may not be open before 1996, and the upper trailhead will not be accessible by road until then.

A short climb over a white pumice levee on the edge of the debris avalanche zone begins the route. Tall peeled posts mark the way across open

This dike, seen along Boundary Trail, intruded into older volcanic strata about 30 million years ago.

flats. In a short distance the path crosses beneath the base of a dike that intruded into older volcanic strata about 30 million years ago. The dike, running uphill like a small version of the Great Wall of China, is about 3 feet wide, with a 10- to 30-foot-high sheer vertical wall on its north side. The south side, once equally dramatic, is now banked up to its top with debris avalanche deposits. This is one of several similar lava dikes in the area that probably were conduits for local volcanic flows.

The trail joins an abandoned logging road at the west end of Johnston Ridge then heads east, skirting the upper edge of a 1,000-foot-high blast-scoured bank overlooking the debris avalanche and the Pumice Plain and peering into the throat of the new crater. If you are nervous about vertical exposure, a few spots along this path may cause you concern.

This trail segment ends at a switchback in Highway 504, which can be walked 0.7 miles east to the Johnston Ridge Observatory. Boundary Trail #1 continues east from the observatory.

Natural Arch

Hiking trail to a natural arch

Trail: Boundary Trail #1
Rating: (M, X); hikers from the west, mountain bicycles from the east to junction with #230
Distance: From Johnston Ridge Observatory 4.6 miles. Until Highway 504 is completed to the observatory, the arch will not be accessible from the west via #1. Until then, alternate routes are (1) 10.4 miles from Coldwater Ridge Visitor Center via #211 and #230, or (2) 9 miles from Windy Ridge via #207 and #1, and 9.8 miles from the Norway Pass trailhead via #1.
Elevation: Johnston Ridge Observatory 4,210 feet, Natural Arch 5,200 feet
Maps: USGS Spirit Lake West, USFS Mount St. Helens Ranger District
Driving Directions: Follow directions to Boundary Trail (West End), preceding. Once the remainder of Highway 504 is opened to the Johnston Ridge Observatory in 1996, continue east on the highway for another 6.7 miles to the observatory.

Prior to the 1980 eruption this striking arch was unknown, hidden by thick forest on a steep hillside. The blast wave quickly changed this, as it scoured the slopes down to bare earth. The rock formation itself, fortunately, withstood the blast. When Boundary Trail was reestablished, it was rerouted to run directly through the newly discovered arch. Hikers heading east through the arch will discover the snowy summit of Mount Adams on the horizon to the right; those going west will find the skyline filled by Johnston Ridge.

Until the road to the Johnston Ridge Observatory is completed and opened to the public, reaching the arch is a demanding task, involving a round trip of 16 to 20 miles from the nearest accessible trailhead,

A natural arch, on Boundary Trail, frames Mount Adams.

depending on your route. Once the observatory road is opened, hikers can drive to the road-end parking lot and head east on Boundary Trail #1. The trail initially follows the rim of an 800-foot-high cliff with a dramatic, unimpeded view across the desert-like Pumice Plain and up the lava flow of the Sasquatch Steps into the heart of the crater.

In 1.5 miles the route bends south across the face of a wall, 200 feet straight up and 450 feet straight down—not the place for anyone with acrophobia! The worst is past in about 500 yards, as the path rounds the nose of a ridge and descends easier slopes to the junction with Trail #207. This latter trail heads south across the barren Pumice Plain to Windy Ridge. Here also is an excellent view of the huge debris dam that reshaped Spirit Lake and raised its water level by more than 180 feet.

An easy uphill grade crosses portions of the debris avalanche that spilled over Johnston Ridge, then skirts the northwest base of Harry's Ridge. The ridge is named for Harry Glicken, a geologist who monitored volcanic activity on St. Helens. Glicken was killed in a volcanic eruption in Japan in 1991.

The trail reaches a saddle near the head of South Coldwater Creek with views east across an arm of Spirit Lake, still partly covered by rafts of logs blown down in the 1980 eruption. Here, too, is the junction with Trail #1E, which swings south atop Harry's Ridge to what is possibly the best view of the crater and Spirit Lake within the entire monument. From the crest of this narrow ridge, slopes on one side drop abruptly for over 1,000 feet to the west shore of Spirit Lake. An equally high, near-vertical cliff on the west side of the ridge drops away to the fractured hummocks that merge southward to the powder-white Pumice Plain. The maze of erosion channels and blocks of debris gradually rises to the foot of the Sasquatch Steps, a collection of blast debris and lumps of pumice that climbs steeply to the plateau in the heart of the crater. Here, swirling steam from vents marks where the future summit dome is slowly building.

From the saddle a switchback eases the uphill climb through the blast-downed forest, then the trail snakes up and around a knob built 30 million years ago from andesite volcanic flows to the first views of St. Helens Lake. On the northwest shoulder of this hill is the arch, the target of our trip. A stream flowing down the steep hillside wore through less-resistant strata, leaving behind a fascinating arch of more solid material.

Return the way you came, or continue on for more extended hikes. Beyond the arch the route reaches the junction with Trail #230 in 0.3 miles.

Norway Pass Loop

Hiking trail to views of Spirit Lake, the Spirit Lake pluton, the crater of Mount St. Helens, and a basalt monolith

Trails: Boundary Trail #1, Independence Pass Trail #227, Independence Ridge Trail #227A
Rating: (M); hikers, mountain bicycles
Distance: Norway Pass 2.2 miles, loop trail 4 miles
Elevation: FR 26 trailhead 3,640 feet, Norway Pass 4,508 feet
Maps: USGS Spirit Lake East, USFS Mount St. Helens Ranger District
Driving Directions: From I-5 take Exit 68 and follow Highway 12E for 50.4 miles to Randle. Turn south on Highway 131, and in 1 mile bear west on FR 25 (2-lane paved). Continue south for 19.6 miles, then turn west on FR 99 (2-lane paved). In 9.3 miles turn north on FR 26 (1-lane paved with turnouts), and in 0.9 mile arrive at the Norway Pass trailhead for Boundary Trail #1.

Hike through the blast zone, with its countless acres of flattened forest splayed out to the northeast and spectacular seasonal wildflower displays. Final destination is an airy overlook of Spirit Lake and beyond into the throat of St. Helens. Before heading up, fill your canteen at a pump at the trailhead parking lot off FR 26 because the trail can be hot and dry.

The trail makes a few long switchbacks in its gradual, but ceaseless, ascent of east slopes. Huckleberries, pearly everlasting, and fireweed, the first plants to regenerate here, swath the slopes. Rangy fireweed bearing cotton-like tufts in fall bears little resemblance to the slender spikes that paint the hillside with brilliant purple in late spring and summer. Scattered young trees, now 6 to 8 feet tall, were seedlings protected from the 1980 blast by 5 feet of snow. To the east, Mount Adams rises above the denuded shoreline of Meta Lake.

At a switchback about 1 mile up the slope is a junction with Trail #227A. This spur winds south up a very steep face and in 1.3 miles meets Independence Pass Trail #227 at the ridge crest 0.8 mile north of Independence Pass. This route permits a 4-mile loop trip to Norway Pass via Trails #1, #227, and #227A. Prior to the recent completion of #227A, a hike of Trails #1 and #227 required transportation between the Independence Pass and the Norway Pass trailheads.

Above the switchback Trail #1 rounds a knob, with views north to

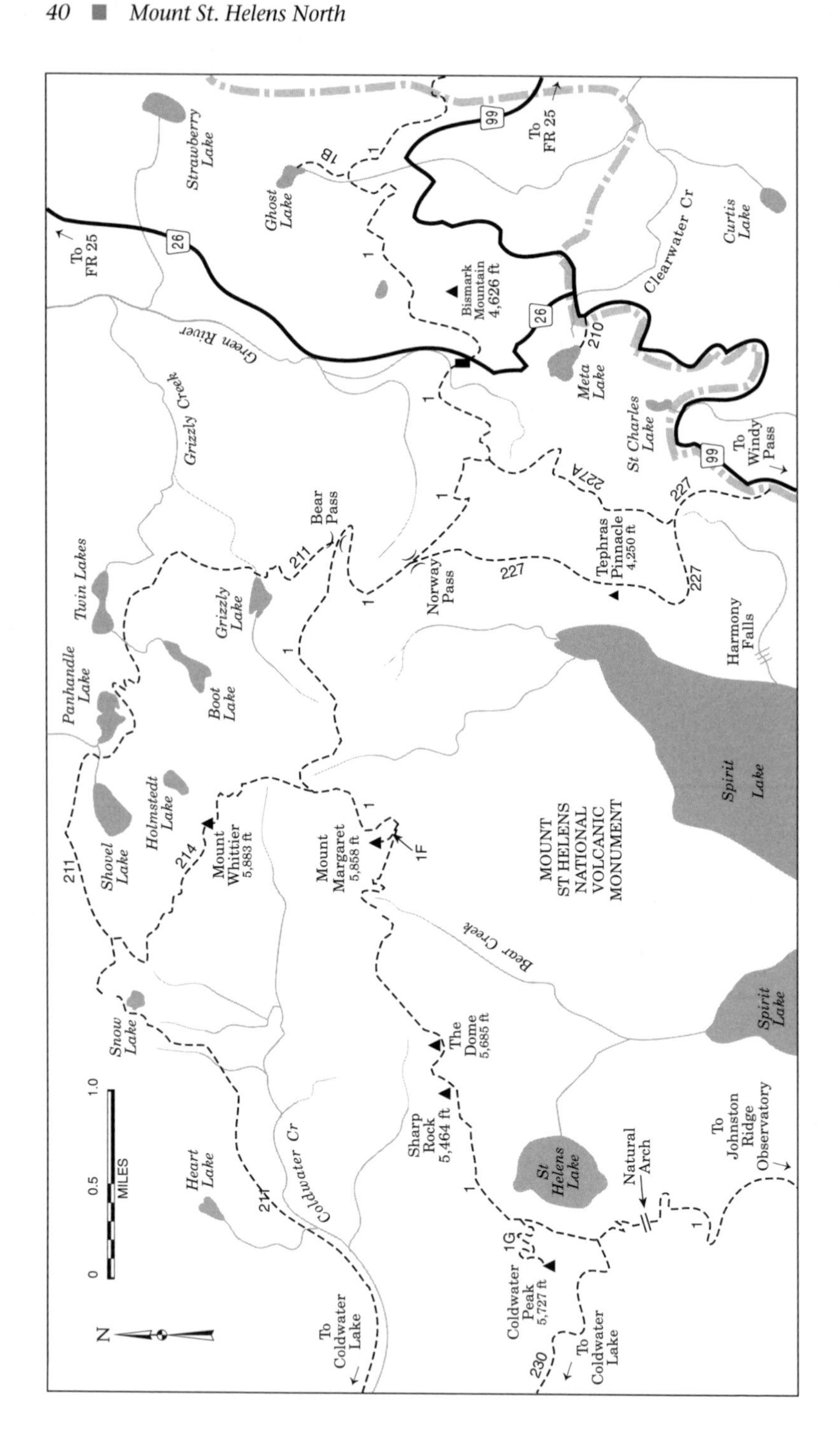
To FR 25
26
Strawberry Lake
Ghost Lake
1B
99
To FR 25
Curtis Lake
Clearwater Cr
Bismark Mountain 4,626 ft
Green River
Meta Lake
210
St Charles Lake
To Windy Pass
Grizzly Creek
Bear Pass
211
Twin Lakes
Grizzly Lake
Norway Pass
1
227
227A
Tephras Pinnacle 4,250 ft
Harmony Falls
Panhandle Lake
Boot Lake
Spirit Lake
Holmstedt Lake
Shovel Lake
214
Mount Whittier 5,883 ft
Mount Margaret 5,858 ft
1F
MOUNT ST HELENS NATIONAL VOLCANIC MONUMENT
Bear Creek
Snow Lake
The Dome 5,685 ft
Sharp Rock 5,464 ft
St Helens Lake
Natural Arch
To Johnston Ridge Observatory
Heart Lake
Coldwater Cr
1G
Coldwater Peak 5,727 ft
To Coldwater Lake
230
0 0.5 1.0
MILES
N

Rainier. In spring, avalanche lilies here bear record numbers of blossoms. This may be caused by a change in nutrients due to the volcanic ash. The path then begins a long down, then up, traverse high above a headwater tributary of the Green River. Granite slabs on the slope to the north are outcrops of the Spirit Lake pluton. The Chicago Mine was located deep in the valley below; patented mining claims still exist downstream.

In 2.2 miles the trail arrives at Norway Pass. At the foot of the deep drainage to the south is the eastern arm of Spirit Lake. For years after the 1980 eruption the lake was almost totally covered with a raft of blown-down logs. Over the past 15 years many became waterlogged and sank, leaving only a wide fringe of logs floating along the shoreline. The open waters of the lake are ultramarine blue, reflecting the truncated summit of St. Helens, whose open crater maw fills the horizon. The steep shoreline of the lake is still rimmed by a bare bank, up to 500 feet high in places, that was stripped of vegetation by the surge of waves triggered when the debris avalanche dammed the south end of the lake and slammed its

The Spirit Lake Pluton

The peaks and ridges north of Spirit Lake are exposures of the Spirit Lake pluton, a large quartz diorite to granite intrusion emplaced in overlying rock between 23 and 18 million years ago. The southern edge of the pluton runs west from Bismarck Mountain, then along the ridge on the south side of Coldwater Creek. Its western extent is along the east side of the Miners Creek drainage, the north side runs on a line through Black Mountain and Goat Mountain, and the east side roughly follows Strawberry Ridge to Bismarck Mountain.

Mining exploration on the fringes of the pluton started in the 1890s, and claims were soon staked near the northeast corner of Spirit Lake; these were later developed as the Sweden and Lang mines. Other claims were filed at the Chicago Mine, just east of Norway Pass. Prospecting was done, and claims were filed along Miners Creek, but little active mining took place here. The Commonwealth Mine was opened on the southeast slopes of Bismarck Mountain. The Green River drainage probably saw the most activity, with operations at the Independence, Minnie Lee, Black Prince, Last Hope, and Polar Star mines, all located along the river on the southwest side of Goat Mountain and Vanson Peak. The Samson claims southwest of Ryan Lake were developed as the Golconda Mine. Ores found were mainly copper-bearing with only small amounts of gold, silver, and lead. Difficult access, high transportation costs, difficulty working the veins, and problems refining ores that were found meant meager profits. As a result, most of the mining in the region shut down by the 1920s.

Spirit Lake, as seen here from Norway Pass in 1994, reflects the devastated summit of Mount St. Helens. Logs on the lake surface are remains of the 1980 blast.

waters abruptly to the northeast, like water sloshing in a dishpan.

To complete the loop trail, follow Trail #227 as it climbs steeply to the south out of Norway Pass, gaining a bench above cliffs that drop away to Spirit Lake. You may spot fragments of youth camps from the shores of the former Spirit Lake that were blown nearly 1,200 feet uphill by the force of the eruption. In about 2 miles the trail passes Tephra's Pinnacle, a 30-foot-high pillar exposed by the blast. At 2.7 miles the path meets Trail #227A, described earlier, at a saddle just beyond Harmony Creek. Follow this trail north to its junction with Boundary Trail #1.

Mount Margaret Ridge

Hiking trail along the south edge of the Spirit Lake pluton with views of the crater of Mount St. Helens

Trails: Boundary Trail #1, Mount Margaret Trail #1F
Rating: (M); hikers, mountain bicycles
Distance: Norway Pass trailhead to Mount Margaret trailhead 4.9 miles, #1F 0.1 mile
Elevation: Norway Pass trailhead 3,700 feet, Mount Margaret 5,858 feet
Maps: USGS Spirit Lake East, Spirit Lake West; USFS Mount St. Helens Ranger District
Driving Directions: Follow directions to Norway Pass, preceding.

This high, ridge-crest trek amid cliffs and pinnacles offers endless views of Spirit Lake, St. Helens, and the deep, rugged ridges and pocket lakes in the remote north end of the monument. See Norway Pass Loop, preceding, for a description of the trail to Norway Pass. From the pass the route rounds the head of the drainage to the north and takes a leisurely switchback near Bear Pass, where it passes the trailhead of Lakes Trail #211 (to Grizzly, Obscurity, Boot, Panhandle, Shovel, and Snow lakes). It continues relentlessly uphill for another 1.2 miles to a saddle. Looking south, the northeast tip of Spirit Lake is still visible, and St. Helens dominates the scene. Slopes to the north drop away precipitously to the upper Grizzly Creek drainage with its coating of blown-down timber. On the opposite side of the drainage, a granite wall—an exposure of the Spirit Lake pluton—soars 450 feet above Grizzly Lake; other rock pinnacles line the ridge to its southwest. Acres of leveled forest, attesting to the colossal impact of the blast wave, can be seen extending for more than 2 miles to the southern slopes of Goat Mountain. Mount Rainier sleeps (for the moment) on the horizon.

The route diagonals gradually up open slopes to the west where beargrass, covered by snow in May of 1980, survived the eruption. The path then rounds a bald ridge nose, where the ramparts of castle-like Mount Margaret come into view beyond an intervening ridge. With the worst of the uphill drag behind, the open rocky ridge crest and its ever-changing perspectives of cliffs, pinnacles, pocket lakes, and the omnipresent crater lure the traveler on until tired muscles and reminders of the length of the trip back to the trailhead demand a reluctant turnaround.

A gentle grade atop a broad, bare ridge reveals breathtaking views down 900-foot-high cliffs to Grizzly Lake on one side and drainage headwalls dropping to Spirit Lake on the other. The way meanders east across a broad flat at a T-junction of ridges, where Mount Margaret and intervening ridgetop spires come into full view. The junction with Trail #214 is reached 1.3 miles from Bear Pass. Before planning to add Trail #214 to future itineraries, take this opportunity to visually follow it to the northwest. For more than a mile, the 30-inch-wide tread carves across the southwest face of

Avalanche lilies along Trail #1, en route to Mount Margaret, bear unusual numbers of blossoms.

Mount Whittier, 50 to 100 feet below the ridge crest. Below the narrow trail a rocky, treeless cliff drops more than 1,400 vertical feet to the headwaters of Coldwater Creek—not for the nervous!

After circling the base of ridge-top spires, the route traverses easy slopes broken by rock outcrops below the summit of Mount Margaret. A short spur, Trail #1F, branches off, heading to the top of the mountain. The summit may have been used for a temporary fire observation site in the 1930s, but no permanent structure was ever erected here.

For those who wish to venture farther, the trail continues west, working carefully along a short, narrow section of crest with ever-expanding views of the south end of Spirit Lake and the debris dam that reshaped it. The ridge top once again broadens and the path weaves gently back and forth among crest outcroppings of the Spirit Lake pluton. In another mile is The Dome, a steep-walled, 200-foot-high knob with outrigger ridges to the northwest and southwest, each topped with a cockscomb of short, granite pinnacles. As the route skirts the south side of the peak and its southwest ridge, St. Helens Lake appears, tucked in a glacier-carved pocket on the south side of the ridge.

The trail gradually descends to the southwest along the upper rim of this cirque, where it meets spur Trail #1G to the top of Coldwater Peak. Switchbacks thread up through cliff bands on the east side of the peak to the former fire lookout site. Here are spectacular views of Spirit Lake and the gaping north side of St. Helens, as well as Minnie Peak and the rocky needles on the ridge north of Coldwater Creek. The lookout atop Coldwater Peak, built in 1935, was destroyed in 1968 by the Forest Service—they should have waited another twelve years and let the mountain do the job for them.

For those who are in prime shape, and who have prearranged transportation at the opposite trailhead, the entire 14.4 miles from the Norway Pass trailhead to the Johnston Ridge Observatory can be made in one very long day. An overnight trip will permit a more leisurely pace, and along with it the opportunity to sample other new backcountry trails, but be aware that the entire area is without tree-cover protection from weather, and no water is available en route.

2

Mount St. Helens South

Although the south side of Mount St. Helens largely escaped the blast devastation and pyroclastic flows that so dramatically altered the north side of the peak during the May 1980 eruption, it was not left untouched. Lahars and ash deposits covered most of the slopes. On the west side of the mountain, the fiery blast of the eruption leveled timber on the north rim of the South Fork of the Toutle River, and massive lahars surged down the river channel and adjoining Sheep Canyon.

To the southeast, the eruption surged up and over the newly formed crater rim, breaching it near the head of the pre-eruption Shoestring Glacier. Portions of the collapsing summit, as well as the upper part of the glacier, formed a scouring flood of mud, rock, and ice that swept down the southeast side of the mountain, sending millions of tons of debris surging down the channels of Swift Creek, Ape Canyon, and the Muddy River. As the immense, abrasive mass of mud, rocky basalt boulders, and glacial ice swept onward through the narrow channels of the creeks and rivers, it scoured hillsides bare for as much as 300 to 500 feet above the old river beds. In broader spots, the flow laid down a massive plain of debris, as much as 30 feet thick. Everything in the path of these lahars was flattened or buried. Trees still standing along the upper margins of the flows show the scars of stripped bark as high as 10 feet on their upstream sides.

Bomb-like craters in the South Fork of the Toutle River lahar are believed to have been caused by the explosion of huge chunks of glacier ice.

Today, some 15 years later, the rivers that originally drained these valleys have started their slow, but inexorable, cutting through the lahar deposits, forming steep-walled banks, 10 feet or more high, that constantly crumble under the relentless erosion of the fast-flowing waters.

Roads into the Mount St. Helens National Volcanic Monument from

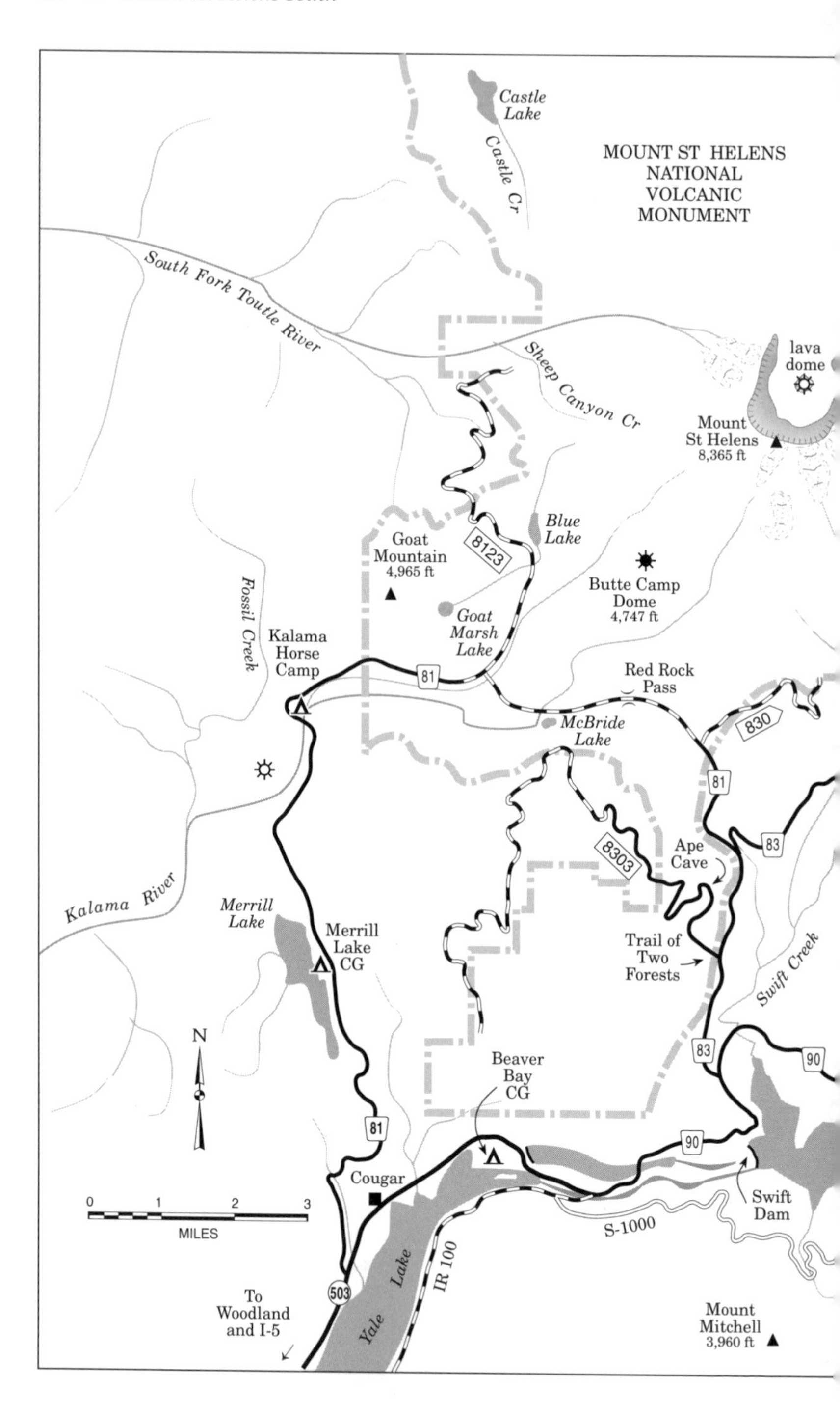

Castle Lake
Castle Cr
MOUNT ST HELENS NATIONAL VOLCANIC MONUMENT
South Fork Toutle River
Sheep Canyon Cr
lava dome
Mount St Helens 8,365 ft
Blue Lake
8123
Goat Mountain 4,965 ft
Butte Camp Dome 4,747 ft
Fossil Creek
Goat Marsh Lake
Kalama Horse Camp
81
Red Rock Pass
McBride Lake
830
81
83
8303
Ape Cave
Kalama River
Merrill Lake
Merrill Lake CG
Trail of Two Forests
Swift Creek
N
83
90
Beaver Bay CG
81
90
Cougar
0 1 2 3
MILES
S-1000
Swift Dam
IR 100
Yale Lake
503
To Woodland and I-5
Mount Mitchell 3,960 ft

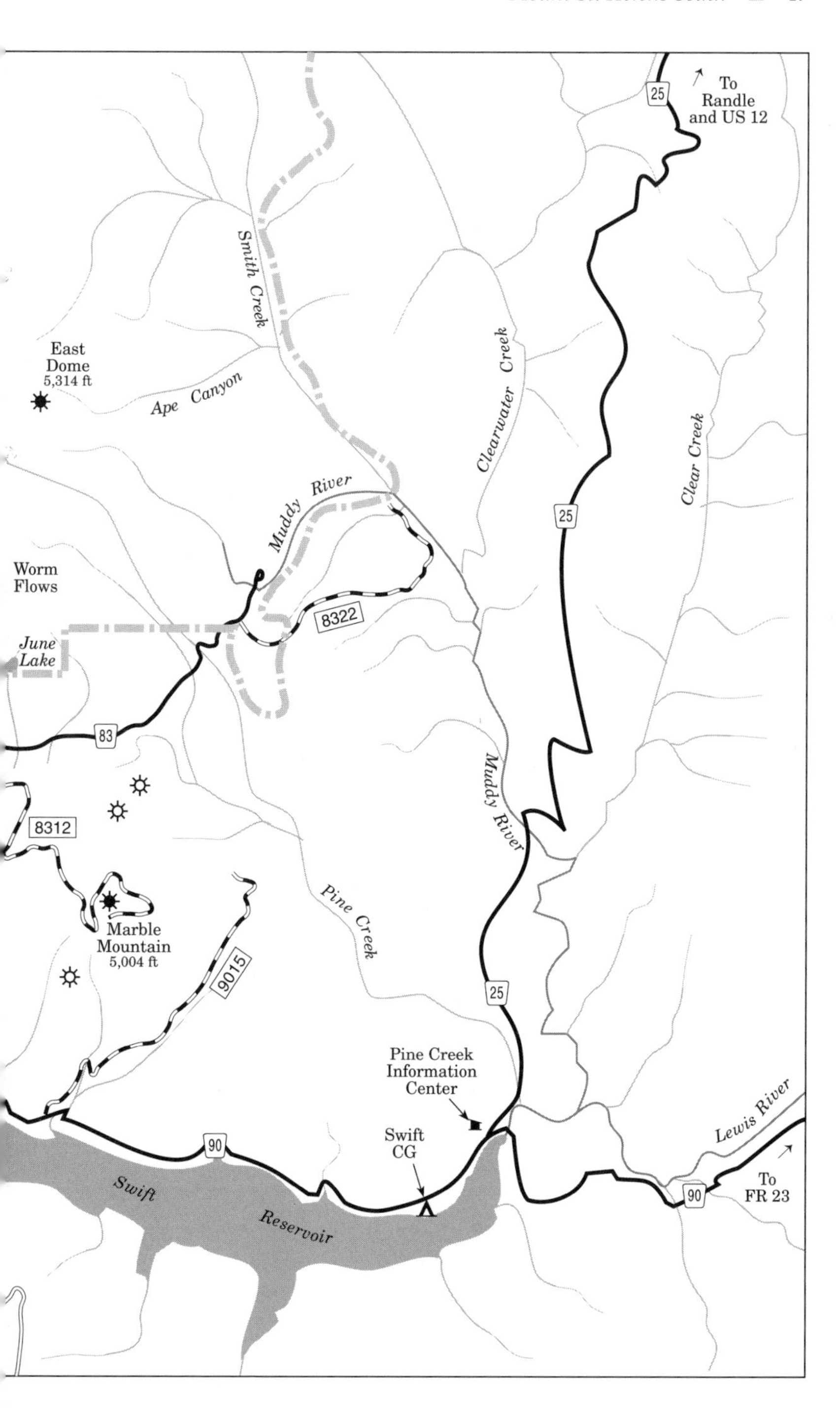
To Randle and US 12
25
Smith Creek
East Dome 5,314 ft
Ape Canyon
Clearwater Creek
Clear Creek
Muddy River
25
Worm Flows
8322
June Lake
83
8312
Muddy River
Marble Mountain 5,004 ft
Pine Creek
9015
25
Pine Creek Information Center
Swift CG
90
Lewis River
To FR 23
90
Swift
Reservoir

the south all originate from Highway 503-Spur or FR 90, which thread up the lower Lewis River drainage, skirting the north rims of Lake Merwin, Yale Lake, and Swift Reservoir. These three lakes were created by dams built by Pacific Power and Light. On the south side of the lakes most of the land, which is owned by the state DNR and private timber companies, has

The Cave Basalt Lava Flows

The Mount St. Helens basalt flows are known to hold sixty lava tube caves and probably others not yet discovered. These are of relatively recent origin, around 2,000 years old. Their source, a vent above the 4,800-foot-level of St. Helens, has been obscured by more recent lava and mud flows. This Cave Basalt lava flow split around both sides of Green Mountain and finally stopped near the north bank of the Lewis River. A chemically similar basalt flow found about 15 miles to the southwest in the Kalama River valley near Lake Merrill may either be part of the Cave Basalt flow or a contemporaneous flow.

The lava tubes in the Mount St. Helens–Mount Adams area contain an amazing variety of vegetation, mammals, and invertebrates. Mosses and a few patches of lichen surround most cave entrances. In the dim twilight zone inside the mouth more than twenty-four species of mosses have been identified. Lush fern gardens drape some of the entrances and liverworts and lichen are found in the semi-darkness of the interiors.

These secluded caves attract residents such as Douglas squirrels, deer mice, woodrats, coyotes, salamanders, and (as expected) bats. Bones of larger mammals such as beavers, bobcats, porcupines, and black-tailed deer have been found in the caves, but they were probably accidental visitors who became entrapped.

And, true to what you've probably always thought about caves, the most populous and varied creatures are the creepy crawlies. Identified invertebrate species include tiny mollusks and crustaceans, more than seventy-five kinds of insects, and over fifty kinds of spiders, as well as an assortment of millipedes, worms, and some unclassified species. Among this wildlife found in the cave systems are nine rare and vulnerable species, among them Townsend's big-eared bat, larch mountain salamander, and blind white copepods.

These geological and biological treasures are unique and irreplaceable. Unfortunately, even the most careful of visitors cause incidental damage to the caves and their inhabitants, and the results of deliberate vandalism can be devastating. Many of the caves in this flow have already seen heavy use by humans. Except for Ape Cave, the Forest Service discourages entry into any of them and will not give directions to their specific locations.

been heavily logged. A side benefit (if it can be construed as such) of this massive clearcutting is that the underlying geological structure and outcrop cliffs previously hidden by timber are now exposed, and their origins are more readily identifiable.

Two major routes branching from FR 90 lead into the south part of the mountain. The westernmost, FR 81, heads north from Highway 503-Spur, 0.3 mile west of Cougar, and winds uphill to Merrill Lake. The lake is believed to have been formed some 1,900 years ago when a north-flowing tributary of the Kalama River was dammed by Cave Basalt flows from St. Helens. The road traces the eastern shoreline at the base of steep, sometimes wooded, sometimes logged, hillside. Midway up the lakeshore is a small DNR campground. Continuing north up the widening valley floor to the Kalama River drainage, FR 81 arrives at Kalama Horse Camp in 1.3 miles. The campground is reserved for horse campers; in early fall other campers will rarely find it occupied.

Mount Mitchell

Hiking trail to the top of a granodiorite intrusion

Trail: Mitchell Peak Trail (DNR)
Rating: (D) from the north, (M) from the south; hikers, saddle and pack stock
Distance: Trailhead to the summit from the north 2.2 miles, from the south 10.3 miles
Elevation: North trailhead 1,940 feet, south trailhead 1,680 feet, summit 3,960 feet
Maps: USGS Mt. Mitchell, Siouxon Peak; USFS Mount St. Helens and Wind River ranger districts
Driving Directions: From the north, turn south from Highway 503-Spur onto IP 100, 0.5 mile east of the Beaver Bay Campground road at Yale Lake. After crossing the Lewis River, turn south on road S-1000 (1½-lane gravel), and follow it east for 4 miles to an unnumbered spur that leads 0.2 mile south to the trailhead. From the south, head east from Highway 503 at Chelatchie on Healy Road. In 5.7 miles turn north on road S-1000; follow it (1-lane paved) for 6.2 miles, then turn north on S-2000 (1-lane gravel). The trail is on the east side of the road in 1.8 miles.

Looking south from the Swift Reservoir Overlook on FR 90, Mount Mitchell dominates the skyline. Its three sharp fangs of rock rim a glacier-cut basin that lies on the mountain's northwest side. These rock snags are the erosion-exposed heart of a mass of granodiorite that, some 20 million years ago, intruded into overlying layers of andesite and basalt. A fire lookout, built in 1940 on the highest of the these pinnacles, was destroyed in 1970.

Two routes lead to the summit; the one from the north is short, rough, and quite steep; the route from the south is in better condition and is much

Mount Mitchell, seen here from near Swift Reservoir, is a granodiorite intrusion dating from 20 million years ago.

more gradual, except for one steep section, but it is four times longer—take your pick.

From the north, depending on its condition, drive or hike the spur from road S-1000 for 0.2 mile to the old trailhead. This unmaintained trail, which leads to the saddle east of the summit, can still be hiked, although it is strenuous (1,000 feet of elevation gain in a mile!) and may require some routefinding. At a saddle in 1.1 miles reach the junction with the trail from the south and the beginning of the summit trail.

The route from the south leaves road S-2000 and drops down logged

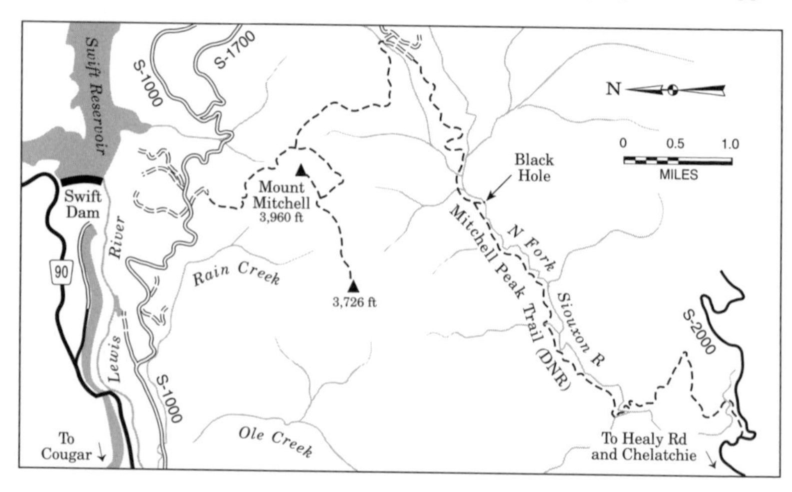

slopes to a footbridge over the North Fork of Siouxon Creek. It then turns east and follows the creek bank upstream, initially on an abandoned logging road, then continuing as a trail. Several side streams that cut the trail are easily forded. In 5 miles the route switchbacks up a finger ridge between forks of a stream to reach another section of abandoned road. In another 0.6 mile, at a switchback at the head of the drainage, the trail resumes and contours a wooded sidehill northwest for 1.8 miles to the saddle junction. From the saddle, the last, strenuous mile of trail gains 1,000 feet as it diagonals up the cliffy southeast face of the mountain, then swings up open slopes on the south rib before bending east for the final thrust to the bald rock summit.

Views here are spectacular, as one would expect of an old lookout site. The shimmering blue of Swift Reservoir fills the bottom of the deep valley to the north, and beyond it forested slopes rise rapidly to the barren, brown, pumice-clad flanks of St. Helens. To the south a green carpet of (yet) uncut timber drops steeply away to the Siouxon Creek drainage. Slopes to the east are another story; privately owned sections of land have been scalped and scarred with a spaghetti of logging roads. Vestiges of forest cover don't reappear until the terrain rises to the distant backbone of the Trapper Creek Wilderness.

Goat Mountain Views and Sheep Canyon

Hiking trail views of blast wave and lahar devastation

Trails: Blast Line Trail #240A, Sheep Canyon Trail #240, Loowit Trail #216, Toutle Trail #238

Rating: #240 (M), #216 (D); hikers, mountain bicycles; saddle and pack stock, llamas on the lower section of #240 to the #238 junction, then south on #238 only

Distance: 5.1 miles

Elevation: Trailhead 3,400 feet, junction of #240 and #216 4,600 feet, South Fork of the Toutle 3,250 feet

Maps: USGS Goat Mountain, Mount St. Helens; USFS Mount St. Helens Ranger District

Driving Directions: Turn north from Highway 503-Spur onto FR 81 (both 2-lane paved), 0.3 miles west of Cougar. Follow FR 81 north for 11.4 miles to FR 8123 (1-lane gravel); continue north on it for 6 miles to the road end.

After passing Kalama Horse Camp, a long, straight section of road focuses views northeast to Goat Mountain and its southwest ridge. Impressive outcrops of dacite lava plug domes, dating from 3.2 million years to 700,000 years ago, form the summit and southeast ridge of Goat Mountain. At 2.3 miles beyond Kalama Horse Camp is the junction with FR 8123. FR 8123 continues north through a thinning pine forest, passing the trailhead for Trail #240 to Blue Lake in another 1.7 miles. The road winds

Sheep Canyon Creek was swept clear of trees by a lahar.

uphill (always take the right-hand fork at junctions) and eventually arrives at a parking lot. Open hillsides along the way offer excellent views of the north side of Goat Mountain and its companion dacite volcanic plugs on the ridge to the east.

From the road-end parking lot, a 0.2-mile path, Trail #240A, leads northwest to an open knob, with views northeast into the South Fork of the Toutle and Sheep Canyon. However, this glimpse can only whet the appetite for the real feel of the results of volcanism that lie ahead to the east along the trail system.

Sheep Canyon Trail #240 heads east from the parking lot through dense eye-high brush. In fall, white and yellow blossoms of pearly everlasting and Martindale's lomatium fringe the trail. A stately stand of old-growth fir is soon entered;

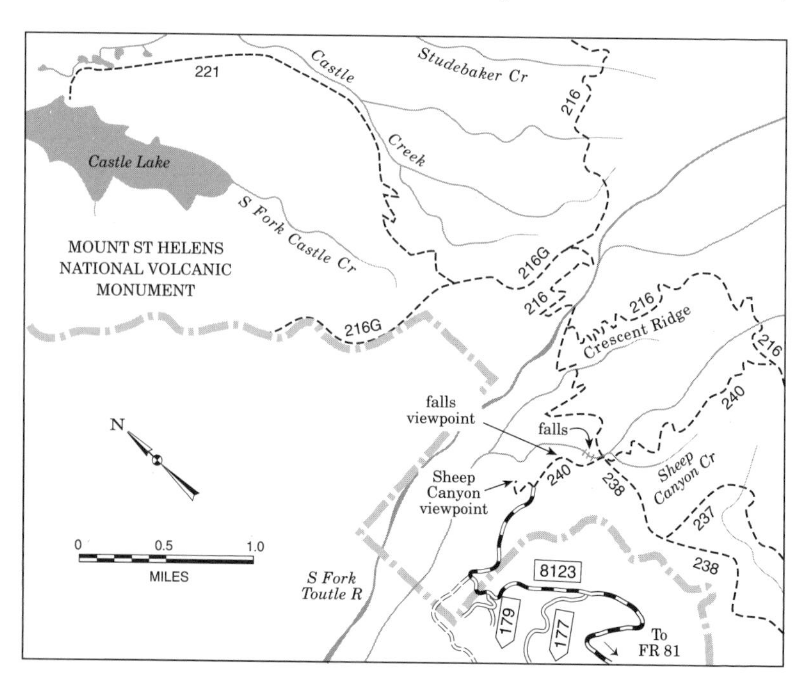

through the trees are glimpses west of denuded slopes of private land that was salvage-logged after the 1980 eruption. Farther along the trail, the monument boundary can be seen clearly demarked by a line between bare, logged slopes and dramatic blow-down from the Mount St. Helens eruption.

In 0.6 mile a bare, tephra-covered rock outcrop provides a terrific view of the Sheep Canyon Creek waterfall. The creek drops over an erosion-resistant basalt-flow lip and falls free in a lovely ribbon of water for about 75 feet, then continues downstream in several lesser cascades as it retreats into the dark, narrow cleft of the lower canyon.

From the viewpoint, the trail climbs a short distance to a bridge upstream from the falls; here it intersects with Toutle Trail #238 from Blue Lake, which crosses the bridge. For the easiest and shortest path to the impressive lahar plain of the South Fork of the Toutle River, take Trail #238 left (north) from this point and return via the same route. (See South Fork of the Toutle River Lahar, following.)

Those with a passion for more exercise and scenery can make a loop trip by continuing steeply uphill on Trail #240 for another 1.6 miles to Trail #216. Follow #216 as it first heads north, then snakes west down Crescent Ridge to a junction with Trail #238. Return via Trail #238. This loop totals 6 miles and involves a gain of 1,000 feet to Trail #216, a loss of 1,450 feet to Trail #238, and a gain of 350 feet back to the start point.

The Trail #216 segment of the loop earns its Difficult rating as it switchbacks abruptly down Crescent Ridge along the southwest fringe of the blast zone. At the start of this descent, a view rock at 4,700 feet provides a breathtaking panoramic view of the deep South Fork Canyon. The path swings through areas denuded by the eruption and past flattened pumice-sanded logs, stands of skeletal trees whose foliage was seared off by the heat from the eruption, and green, untouched forest. The amputated summit of the volcano looms against the skyline.

South Fork of the Toutle River Lahar

Hiking trail to a lahar

Trails: Sheep Canyon Trail #240, Toutle Trail #238
Rating: (M); hikers, mountain bicycles; saddle and pack stock, llamas on #240 below the junction with #238
Distance: 2.2 miles
Elevation: Trailhead 3,400 feet, junction with #238 3,600 feet
Maps: USGS Goat Mountain, Mount St. Helens; USFS Mount St. Helens Ranger District
Driving Directions: Follow directions to Goat Mountain Views and Sheep Canyon, preceding.

The entire drainage floor of the South Fork of the Toutle River was scoured by a 20-foot-high wall of melted glacial ice and pumice, with the consistency of wet concrete, that roared down the valley eighteen minutes

The South Fork of the Toutle River cuts a new channel through lahar deposits. Hillsides above the river were stripped of trees by the force of the blast.

after the initial eruption of St. Helens. To view the effects of this catastrophic event, and the valley's ongoing recovery from it, turn north on #238 at the junction of #240 and #238 and cross the bridge over Sheep Canyon Creek. Note that upstream the drainage bottom is still rocky and bare from the sweep of the lahar; fringes of low brush cover are just now starting to creep toward the creek banks.

North of the bridge the trail meanders gradually downhill across forested slopes to a switchback and the first view of the South Fork of the Toutle and the awesome impact of the eruption. Although the barren hillsides on the north side of the river have been salvage-logged up to the monument boundary, those east of the boundary are covered from base to ridge crest with a blanket of bleached logs. This is all that remains of the former forest that in a matter seconds was uprooted, stripped of branches, sanded clean of bark, and laid flat in an east–west direction by the tremendous lateral force of the blast from the eruption. Farther upslope the hillside transforms into a brown featureless bowl, totally devoid of vegetation. Here the power of the pyroclastic flow stripped away all the trees and scoured away the topsoil, then left behind a loose cover of ball-bearing–like tephra. Still farther up the drainage all surface features have been removed, exposing a thick layer-cake cliff composed of several stacked volcanic flows. Most of the individual flows represented in the strata of this face were laid down between 2,200 and 1,700 years ago.

The former canyon carved by the South Fork of the Toutle is now a broad plain formed when the torrent sped down the river valley. The surge-cut walls on the side slopes above the drainage floor mark the maximum depth of the mud flows. The river is slowly cutting a snake-like channel through the lahar plain, exposing in its steep soft banks the texture of the mud flow, ranging from pea-sized tephra to dark table-sized basalt boulders. Note also a pair of holes that resemble bomb craters near the south side of the lahar. These were created when blocks of glacial ice, encapsulated in the hot mud flow, flashed to steam and exploded, leaving these craters to mark the event.

The trail descends to the lahar floor, passing above a marshy area created when the mud flows blocked the stream draining down the south side of Crescent Ridge. The truncated summit of Mount St. Helens looms above, a steam cloud floating over its rim—raising apprehension about the next explosion of this turbulent giant.

Castle Lake

Hiking trail to a lake created by the debris avalanche

Trails: Sheep Canyon Trail #240, Toutle Trail #238, Loowit Trail #216, Castle Ridge Trail #216G, Castle Lake Trail #221
Rating: (M); hikers, mountain bicycles
Distance: 8.3 miles
Elevation: Trailhead 3,400 feet, junction with #238 3,600 feet, junction with #216 3,250 feet, junction with #216G 4,000 feet, Castle Lake 2,510 feet
Maps: USGS Elk Rock, Goat Mountain, Mount St. Helens; USFS Mount St. Helens Ranger District
Driving Directions: Follow directions to Goat Mountain Views and Sheep Canyon, preceding.

Castle Lake is so near—and yet so far. At the Castle Lake Viewpoint on Highway 504, 35.4 miles east of the Mount St. Helens Visitor Center, you catch your first tempting glimpses of it, midway up the south side of the North Fork of the Toutle River drainage. The lake was formed when the debris avalanche from the 1980 eruption blocked the flow of Castle Creek to form the lake. The Corps of Engineers subsequently cut an outlet through the debris dam to relieve pressure from the enclosed waters and prevent breaching of the dam. While the lake is only a mile away from the Hummocks Trail below Coldwater Lake, no feasible bridging of the North Fork of the Toutle River has been found for an approach to the lake from the north.

At present, the main access to Castle Lake is via the chain of Trails #240,

#238, #216, #216G, and #221, a total distance of 8.3 miles. Follow the preceding trip to the South Fork of the Toutle River lahar; at the junction of Trails #238 and #216 at the South Fork of the Toutle, the path fords the river, then switchbacks up through blown-down timber to barren open slopes. A long switchback across the dry sun-baked bowl reaches the ridge crest and the junction with access Trail #216G. This route heads west for 0.8 mile to meet Trail #221; the latter runs north through blown-down timber, then drops steeply through barren, blast-cleared slopes and pyroclastic flows to the southwest rim of the debris avalanche. This debris avalanche is made up of the fractured remnants of the pre-eruption cone of St. Helens. The path traces the base of the bare ridge above the avalanche, then climbs gradually to the debris dam that created the lake. Trout fishing, it is claimed, is great, but the trip is a long, dry slog and definitely a challenging one-day round trip. Camping is permitted; however, there are no formal campsites at the lake.

Butte Camp Dome

Hiking trail through a lava flow to a dacite dome and start of a climbers' route to the summit crater

Trails: Toutle Trail #238, Butte Camp Trail #238A

Rating: (M); #238 hikers, mountain bicycles, saddle and pack stock, llamas, #238A hikers, mountain bicycles; *travel above junction of #238A and #216 open only to climbers with permits*

Distance: To Butte Camp 2.5 miles, to #216 4 miles

Elevation: Redrock Pass trailhead 3,116 feet, Butte Camp 4,033 feet, junction with #216 4,780 feet

Maps: USGS Mount St. Helens, USFS Mount St. Helens Ranger District

Driving Directions: Either of two routes can be used. For the first, follow the directions in Goat Mountain Views and Sheep Canyon, earlier. At the junction of FR 90, FR 8123, and FR 81, continue east on FR 81 (1-lane gravel) for 2.3 miles to Redrock Pass. For the second route, 7.5 miles east of Cougar turn north from FR 90 onto FR 83 (2-lane paved), then in 3.2 miles head west on FR 81 (2-lane paved, then 1-lane gravel) for 2.9 miles to Redrock Pass.

The Butte Camp trail leads to a climbers' route to the summit of Mount St. Helens. Hikers not bound for the summit will still be interested in the geology along the way. Much of what is seen dates from the time when the mountain was in the process of building its picture postcard image, prior to 1980. Butte Camp Dome itself is a dacite lava plug, remains of a thick, 1,000-year-old magma intrusion into the flanks of the mountain.

From the trailhead at Redrock Pass, a pair of switchbacks climb over the lip of a blocky andesite lava flow and wander across a 10-acre moonscape jumble of lava boulders, products of a massive eruption that occurred some 2,000 years ago. Uphill, to the north, the partially wooded protuberance on

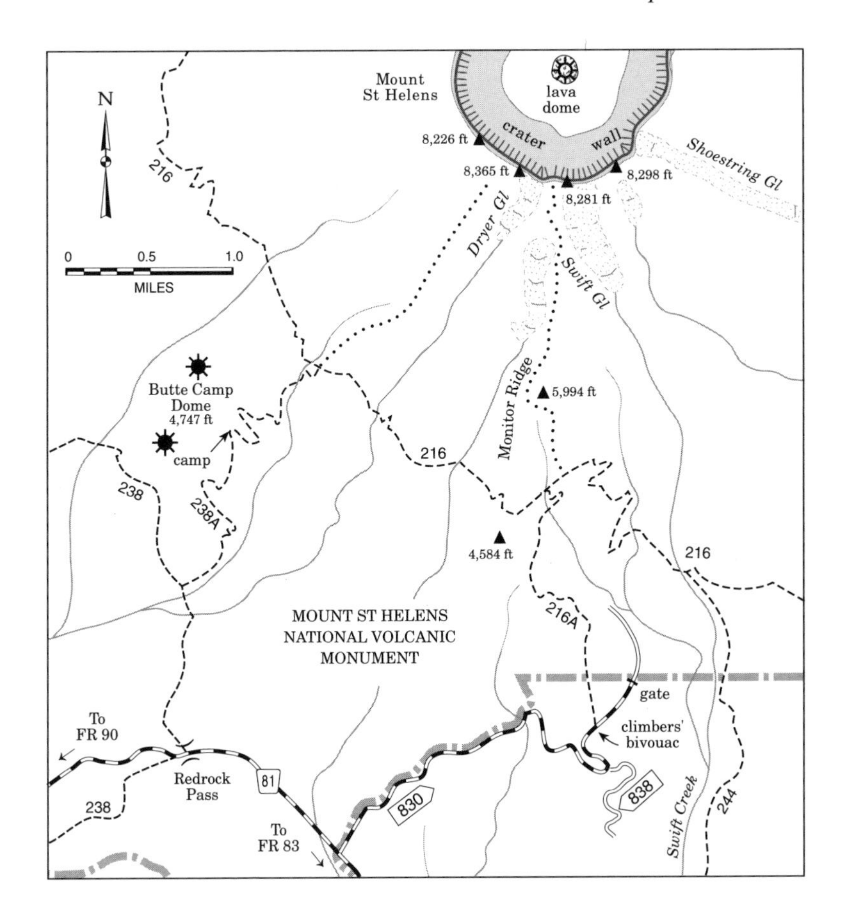

the lower slopes of St. Helens is Butte Camp Dome. Above the dome the slopes of the mountain shift to a barren gray-brown surface of the mud flow and pumice that continue up to the current summit rim.

The trail leaves the bouldery lava flow to cross a beargrass-covered meadow that nurtures a variety of mushroom species that force their way through the pumice soil in fall. The meadow dissolves to open forest of second-growth Douglas-fir, hemlock, and silver fir. Underlying rocks protruding here and there through the tephra of the recent eruption were formed between 500 and 350 years ago when the mountain went through a violent eruptive period that ended with the intrusion of the summit dome of St. Helens. Multiple pyroclastic flows and lahars spilled down the south side of St. Helens during that period.

After a leisurely up and down roll, the path arrives at a Y-intersection. Trail #238 goes left, headed for Blue Lake. Trail #238A, an avenue for mountain climbers headed for the summit rim, branches right. Follow this trail as it switchbacks uphill and passes through a partially exposed flow of 1,900-year-old cave basalt. Watch beside the trail for a quirky, leprechaun-sized lava bridge, the result of solidification of the splashing and spattering of the fluid basalt flow.

At 4,233 feet, tucked in a clearing beneath the flank of Butte Camp Dome, is a climbers' camp. From here, switchbacks grind up the steep sidehill east of the dome to join Trail #216 at timberline. The reward for this final hiking effort is a striking view south to the sharp, glacier-clad summit of Mount Hood. For those with climbing permits, the ever-steepening slopes above lead to the ragged summit rim.

Mount St. Helens Crater Rim

Hiking trail and climbers' route to the summit crater

Trail: Ptarmigan Trail #216A
Rating: #216A (D); hikers, mountain bicycles; *summit climb open only to climbers with permits*
Distance: To #216 2.1 miles, to crater rim 4.6 miles
Elevation: Trailhead (Climber's Bivouac) 3,765 feet, junction with #216 4,680 feet, crater rim 8,365 feet
Maps: USGS Mount St. Helens, USFS Mount St. Helens Ranger District
Driving Directions: 7 miles east of Cougar turn north from FR 90 onto FR 83, then in 2.9 miles head west on FR 81 (all 2-lane paved). In 1.7 miles drive northeast on FR 8100830 (1-lane gravel), and reach Climber's Bivouac in 3 miles.

Two climbing routes are commonly used on the south side of Mount St. Helens, one via Butte Camp Dome, described previously, and the most popular, via Monitor Ridge, accessed by Trail #216A. Any venture above Trail #216 (4,800 feet) requires a climbing permit (available at Jack's Restaurant at Yale, at the junction of Highway 503 and Spur 503). It is recommended that summiters have an ice ax and crampons when snow or ice are still present, be experienced in their use, and carry the Ten Essentials.

From Climber's Bivouac on FR 8100830, Trail #216A first climbs gently north through open forest, then swings west to pick up a lightly wooded ridge leading along the lower edge of the Swift Creek lava flow. This flow resulted from a series of explosive eruptions that occurred between 13,000 and 8,000 years ago. Pyroclastic dacite flows and voluminous deposits of ejected pumice blanketed the south face of St. Helens during this period.

The path switchbacks up the steep headwaters of the Swift Creek drainage, then breaks out of timber and joins Trail #216. The trail trip continues 0.2 mile to where the mountain climb begins. The route through the volcanic boulder fields at the lower end of the Swift Creek flow is marked by wooden stakes; in summer the boot-beaten path is clearly evident.

Steepening slopes covered by slippery tephra and ash from the 1980 eruption make for a dusty, difficult ascent up Monitor Ridge after snow has melted. The 4,000-foot grind up this long, desolate slope is more than

Climbers rest on the rim of the Mount St. Helens crater. Mount Rainier is in the distance.

paid for, however, by the awesome view down the near-vertical 2,000-foot cliffs of the crater wall to the bulge of the dacite lava dome budding from the crater floor—the future summit of St. Helens. Standing on the crater rim (or for safety, a few feet back from it) leaves one straining to imagine the magnitude of the huge series of landslides that removed the top 1,300 feet of the peak and uncorked the gas-rich magma explosion that ripped northward from the crater.

Ape Cave

Scramble in a lava tube cave

Rating: (M); hikers
Elevation: Main entrance 2,115 feet, upper entrance 2,480 feet
Maps: USGS Mount St. Helens, Mt. Mitchell; USFS Mount St. Helens Ranger District
Driving Directions: 7.5 miles east of Cougar turn north from FR 90 onto FR 83, and in 1.8 miles turn west on FR 8303 (all 2-lane paved). Reach the parking lot at Ape Headquarters (the USFS Interpretive Center) and the main entrance to Ape Cave, in 0.9 miles.

Ape Cave—the name conjures up fantasies of meeting the mysterious Sasquatch, the huge ape-like creature said to roam the wild backcountry

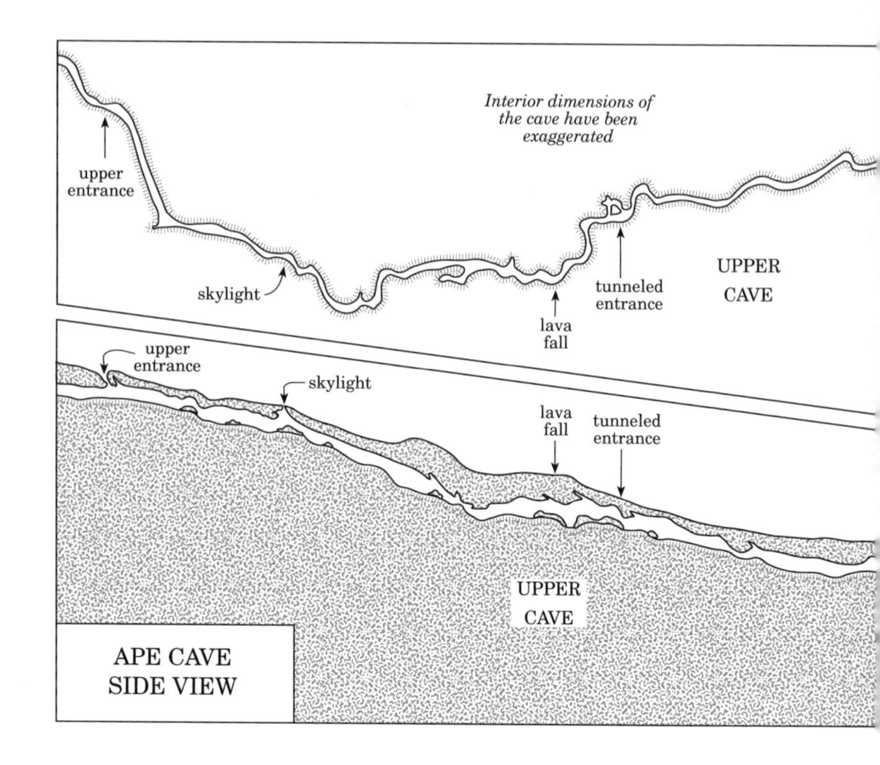

of the South Cascades. Actually, the source of the name is more mundane. The cave was discovered in 1946 when a cat skinner logging in the area came within inches of plunging his tractor into its gaping sink-hole entrance. A local Boy Scout troop called the Mount St. Helens Apes first explored the cave in 1952, and the cave was later named for the group.

Ape Cave was claimed to be the world's longest lava tube when it was first mapped in 1958, and it held that distinction until 1971; since then eight longer caves have been discovered at various spots around the world. At 12,810 feet it is still the longest known lava tube cave on the North American and South American continents.

The cave has no artificial enhancements other than the two entrance staircases—no electric lights, smoothed paths, hand rails, or safety features. As a result, when you choose to explore it, especially its farthest reaches, you are on your own in a pitch black tube, up to ¾ mile from the nearest exit and daylight. The only illumination you will have during your trip are the lights you bring with you. The basalt walls seem to suck up light, and even the brightest lantern soon feels woefully weak. Be sure to have an extra light along (the Forest Service recommends three light sources)—try turning yours off when in the cave and imagine what it would be like trying to grope your way back to the entrance in the dense blackness. According to the Forest Service, a few years ago a young explorer's

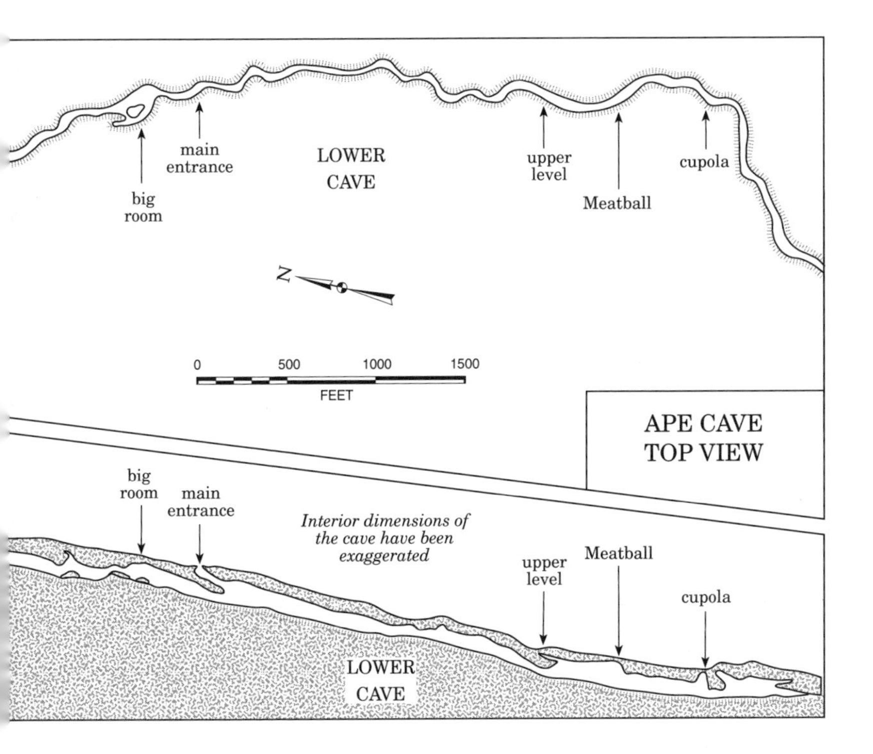

single light source gave out, and he escaped by burning strips of his clothing. He finally emerged from the cave almost naked.

To understand what you will be seeing, take a few minutes to examine the displays at the Forest Service's Ape Headquarters near the main entrance. The center is staffed in summer months, and thirty-minute interpreter-led tours are conducted three times daily on weekdays and six times a day on weekends. Rent a lantern if you don't have one, then walk the well-signed trail to the entrance, where rock steps and a metal ladder descend to the cavern's main tube level. Daylight fades rapidly and you find yourself on the floor of a black-walled tube, roughly 20 to 50 feet wide, and 20 feet below ground at the foot of the entrance staircase.

Warm weather ebbs with the sunlight; cave temperatures average 42 to 48 degrees Fahrenheit, night and day, year-round. Winds or breezes are always present in the cave and, no matter how imperceptible, they lower the effective body temperature even more; sitting or leaning on cold cave walls and floors rapidly conducts body heat away. Warm clothing is needed for visits of more than a few minutes, regardless of the outside temperature.

The steel staircase descends into the middle of the tube; the lower section of the cave lies ahead. Although it varies considerably throughout its length in size, shape, and steepness, most visitors prefer this lower section

Lava Tubes

Lava tubes, trenches, and caves all share a common origin: outpourings of hot (1,600 to 2,000 degrees Fahrenheit), unusually fluid basalt (pahoehoe) from volcano vents. The molten lava seeks the easiest path downhill, often following preexisting canyons carved by erosion through older volcanic flows. The flow of hot basalt further erodes and then fills these canyon troughs. The sides of the lava stream along the canyon walls soon cool and thicken, and the upper surface, in contact with cooler air, crusts over. Shear planes develop in the cooling lava, and soon only the hottest interior portions continue to flow downhill in a sinuous, shifting stream. As this molten interior drains from the congealed walls and ceiling of the lava flow, hollow tubes are left behind.

The resulting tubes act as conduits for further surges of basalt from the source vent for as long as the eruption lasts—which could be months or longer. The hot, viscous streams of lava drain downhill through these tubes of solidified basalt to the ever-expanding toe of the flow.

Occasional heavy spurts of volcanic matter from the source vent sometimes exceed the capacity of the main tube, and smaller branched and braided secondary channels may be created. At times, these surges fracture the ceiling of the tube and break through and run along its top, either thickening its ceiling or creating a new tube stacked on top of the old one. Hot volcanic gases and radiant heat from the flowing lava often remelt the tube walls and ceiling, glazing their surface.

Ceilings, which vary greatly in thickness, are sometimes thickened by later flows. In some cases a roof never forms over the flow channel; the result is a basalt-lined trench rather than a cave. At other times parts of the newly formed roof are so thin and fragile that large windows open up and ceilings collapse, dividing the original tube into separate, shorter caves. Over time ceilings are sometimes worn thin by surface erosion and skylights begin to break through.

of Ape Cave because of its high ceiling and hour-glass shape. Much of the floor is covered with volcanic sand washed in through the main entrance. Other than a few piles of rock that have fallen from the ceiling during the cave's early formation, there is little breakdown debris. If you are careful of your footing, you can go down-tube with relative ease for ¾ mile before the cave squeezes down to a sand-filled crawlway. There is no exit at this end of the cave, and you must retrace your path to regain the surface.

The most famous feature of Ape Cave is the Meatball, found about 2,400 feet below the main entrance. This round ball of lava became wedged in a constriction above the present floor while the lava was still flowing through the cave. More lava balls are stuck together up-tube from the Meatball. These unusual formations are rarely found in other lava tube caves.

The upper portion of the cave, which lies behind you as you descend the entrance stairs, is longer (about 1½ miles) and much rougher. About 350 feet up-tube from the main staircase is one of the largest rooms in the cavern. Note the distinctive flow marks and gutters along its west wall and the 163-foot-long side passage that branches from its northwest corner. Most visitors don't go up-tube beyond this point because of the difficulty in clambering over piles of debris breakdown and scrambling up short lava falls between different floor levels. This portion is also more twisting than the lower section.

Several unique geological features reward the effort of heading up-tube, however. Large rocks, captured by the lava flow and transported downstream

Ape Cave is one of numerous lava tubes in the Cave Basalt flow, which occurred about 20 million years ago.

from the point where they fell from the ceiling, protrude from the floor in places. Later lava surges often didn't fill the entire passage, and deep gutters formed along the side walls. At various spots hot volcanic gasses melted portions of the ceiling to form fragile ribbon stalactites.

A good deal of water splashes into the upper portion of the cave through a pair of skylights or seeps in through thin spots in the ceiling; in one place a small pond occasionally builds up behind a lava dam. About 400 feet from the upper end of the cave, another metal staircase regains the surface through a narrow skylight. An easy 1.5-mile trail leads back to the parking lot. Although the roof of Ape Cave is as much as 30 feet thick in spots, it is generally less than half that thick, and in some places it thins almost to transparency.

Trail of Two Forests

Boardwalk and barrier-free path past tree casts

Rating: (E); barrier-free, hikers
Elevation: 1,890 feet
Maps: USGS Mt. Mitchell, USFS Mount St. Helens Ranger District
Driving Directions: Follow the directions to Ape Cave, preceding. The Trail of Two Forests is on the south side of FR 8303, 0.2 mile from its junction with FR 83 (0.7 mile below Ape Cave).

The Trail of Two Forests, named for the present-day forest and the earlier one that was encapsulated or burned by lava flows, starts at the east side of the parking lot, where an interpretive sign explains what you will see on this easy boardwalk loop. The path weaves through the forest past holes, both horizontal and vertical, where trees were encapsulated by molten lava flows and burned out, forming a rabbit warren of tree-trunk-sized tunnels beneath the forest floor.

Tree casts are fascinating evidence of the Cave Basalt flow.

Midway along the path a short ladder can be descended into a 3-foot-diameter vertical cast that connects to a horizontal one. Venturesome explorers can crawl along the horizontal cast as it heads 20 feet to the north, then takes a 90-degree bend into another 35-foot-long mold. Don't take the short, blocked route to the right. The maze eventually surfaces at the boardwalk path 50 feet to the north. This is a

Tree casts

Lava flows in forested areas burn up most vegetation on contact. However, larger trees sometimes don't reach ignition temperature before the flow lapping around them congeals, encapsulating their trunks in solid basalt. The trees either are burned up, or eventually rot out, leaving a female mold of the trunk in the lava where the tree once stood or laid. Many of these tree casts are found along the margins of cave basalt flows, and some are occasionally linked to the underground lava tubes, forming snug access holes into the caves.

true explorer's thrill for enthusiastic kids, and non-claustrophobic slender adults who don't mind a damp crawl. A flashlight will certainly help allay fears along the tight, 75-foot-long channel. Kids may well demand several cycles through the tree cast tunnel system!

June Lake and the Worm Flows

Hiking trails past lava flows

Trails: June Lake Trail #216B, Loowit Trail #216
Rating: #216B (M), #216 (D); hikers, mountain bicycles
Distance: June Lake 1.6 miles, Worm Flows and Muddy River lahar 4.5 miles
Elevation: Trailhead 2,710 feet, June Lake 3,120 feet, #216 3,400 feet, Muddy River lahar 4,400 feet
Maps: USGS Mount St. Helens, USFS Mount St. Helens Ranger District
Driving Directions: 7.5 miles east of Cougar turn north from FR 90 onto FR 83 (both 2-lane paved), and follow it north, then east for 6.9 miles to FR 8300250 (1-lane paved). Turn north here, and in 0.2 mile reach the trailhead.

A relatively easy trail through open fir wanders uphill for about a mile before reaching a dramatic viewpoint of the huge lava field west of June Lake. This is a part of the massive Worm Flows that erupted from the mountain about 500 years ago; their name comes from the appearance of the flow when viewed from a distance on the south side of the mountain. The trail continues to a flat plain that melds into the pumice beach and shallow green water of June Lake. Across the lake, to the northeast, a pair of 50-foot-high waterfalls drop, ribbon-like, through the steep, forested slopes above the lake.

The lake is certainly objective enough, and the lava flows to the west are adequately impressive, but to further examine the volcanism continue uphill through a brutally steep pair of switchbacks to Trail #216. To either the east or west, this trail skirts the base of the huge blocky andesite mass

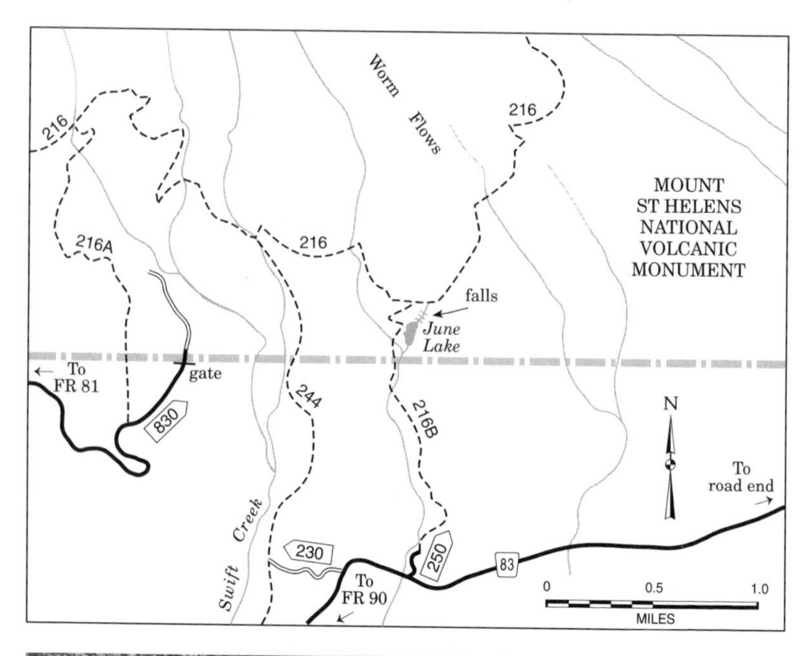

Twin waterfalls feed shallow, green June Lake.

of the Worm Flows. To the east Trail #216 sometimes follows the rough edge of this huge pile of lava blocks but usually stays in the fringe of woods near their base.

In about a mile the path climbs out of the trees and starts a difficult, steep, sidehill ascent across the headwaters of Pine Creek toward the Shoestring Creek and Muddy River lahar below the former Shoestring Glacier. The route becomes increasingly difficult in the loose mud flow surface cut by the stream channels; this may be a good point for a discrete retreat to the more pleasant shores of June Lake.

The Muddy River Canyon

Hiking trails and barrier-free trail through a lahar-swept canyon

Trails: Trails #184, #184A, #184B, and #225
Rating: Barrier-free section of #184 (E), #184A (M), #184 (D, X), #184B (D, X), #225 (D); hikers
Distance: Barrier-free section of #184 0.4 mile, #184A loop 1.2 miles, Lava Canyon Trail 2.6 miles
Elevation: Upper trailhead 2,930 feet, suspension bridge 2,550 feet, lower trailhead 1,575 feet
Maps: USGS Smith Creek Butte, USFS Mount St. Helens Ranger District
Driving Directions: Follow directions for June Lake and the Worm Flows, preceding. Continue east on FR 83 (2-lane paved) past FR 8300250. In another 3.7 miles is the junction with FR 8322. To reach the upper end of the trail, continue east on FR 83 for another 0.8 mile to a road-end parking lot; for the lower end, turn south on FR 8322 (1-lane gravel) and follow it for 4.9 miles as it twists downhill to the start of Trails #225 and #184.

This is one of the most fascinating trails in the monument from a geological, scenic, and adventure perspective, but it is also one of the more challenging. While a round trip technically is only 6 miles, because of its difficulty most persons hiking the full length of the canyon prefer to do it only one way. Start at either end and arrange to be picked up at the other. Although this description describes the trail from top to bottom, most experienced hikers feel that if you intend to do the entire trail, the exposed sections are less nerve-wracking going uphill.

Fortunately, the trail, in general, is divided into increasingly more difficult sections from top to bottom, with the scenery becoming more dramatic in proportion to the difficulty—pick your level of comfort with the trail condition, and be rewarded accordingly. From the top, the easiest piece of trail is a paved barrier-free path that gradually switchbacks down to a viewing platform at river's edge. The scenery is not dramatic, but one certainly gets a feel for the scouring action of the mud flow that surged down the canyon following the 1980 eruption and for its depth, as recorded in bark peeled to a height of 6 feet or more from the uphill side of canyon-edge trees. Downstream on the south side of the river is a glimpse of striking columnar jointing in lava that flowed down the canyon during the Castle Creek eruptive period about 1,900 years ago.

The next in degree of difficulty is the loop completed by Trail #184A that starts just beyond the viewing platform. Hiking the loop clockwise, the path continues north along the west bank of the river, which has cut down 50 feet or more through softer, pale yellow, 30-million-year-old volcanic deposits to expose a wall of columnar-jointed basalt/andesite that rests atop these older layers. The trail at times winds close to the cliff bank with views down into a series of foaming cascades.

In slightly over 0.2 mile, views open to the deep lower canyon, and—shades of Indiana Jones!—a suspension bridge across the abyss comes into view. The west end of the bridge is the start of the most difficult part of the canyon trail and the return point for the more moderate loop. The bridge crosses the canyon over the upper lip of a series of dramatic waterfalls; crossing it is a "both hands on the suspension cables" effort, as it bounces and sways at the slightest shift of weight. On the return trip of the loop, the trail ducks behind a lava outcrop cliff where layers of platy andesite are topped by less viscous pahoehoe lava flows. Next is a climb up a 10-foot-high ladder that tops the lava beds, then the trail heads back to the start point, crossing a bridge over a roaring, 10-foot-high waterfall.

Now for the rest of the trail! From the west end of the suspension bridge the 30-inch-wide path cuts across a bare tephra cliff that drops away steeply for nearly 200 feet to the roaring, waterfall-laden river. At one spot, where a crossing creek makes the tread slippery, a cable handhold has been installed—other than that, sure-footedness and an indifference to heights is a prerequisite for descending the narrow canyon wall. The path descends to the lip of an outcrop of the younger lava flow, where a cliff is conquered by a 30-foot-high metal ladder bolted to the

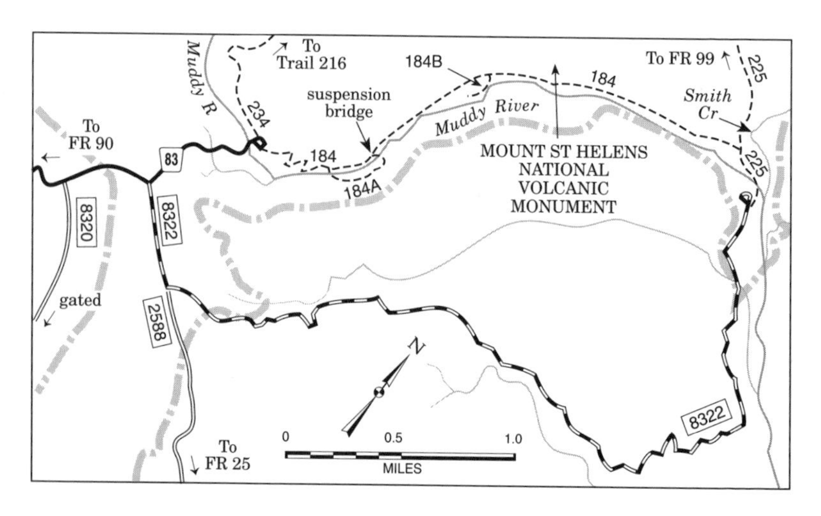

The Muddy River cascades down its canyon. The spectacular cliff of columnar basalt on the right is from the Castle Creek eruptive period.

The suspension bridge across the Muddy River canyon provides an "Indiana Jones-style" thrill.

face. The canyon darkens, as even summer sun rarely climbs high enough to flood into its deep chasm.

Downstream, a spur, Ship Trail #184B, is not for the faint-hearted! In a short 0.2 mile the path climbs steeply up a hundred-foot-high island that diverts the flow of the Muddy River. A 10-foot-high ladder surmounts a rock face; beyond, the path leads up to a ridge-top path, 8 feet wide at most, between encroaching 100-foot-high vertical walls. The trail continues to the south end of the "Ship," ending atop a cliff that overlooks the upper canyon. Ahead is a thrusting 75-foot-high wall of columnar lava flow, stacked atop ancient volcanic deposits. Above its left flank the Muddy River drops down the canyon in a series of several magnificent waterfalls. On the right flank is the ladder you recently used to descend the face of the flow, and on the horizon up-canyon is the west tower of the suspension bridge. To the north, a mud flow–bared hillside shows at least four different layers of volcanic tephra laid down in past eruptions of St. Helens. On the south side of the Ship is a 200-foot-high cliff with fascinating strata bands from old lava flows.

The worst (and best) is almost over, as the trail continues downstream, crosses the top of a 2-foot-wide lava flake, then proceeds across steep hillsides with, at most, 150 feet of vertical exposure. The path drops out of the deep canyon onto a broad lahar plain, where the Muddy River is now cutting a new channel through the soft mud flow deposits. A 0.7-mile hike across the lahar plain, with its thin carpet of moss, miniature arctic lupine, and scattered blocks of pumice and basalt, leads to the lower end of Trail #225. A short hike reaches the 150-foot-long bridge across the Muddy River. Beyond is FR 8322 and the trailhead parking lot. Pheeew!

3

The Columbia River Gorge

As it travels west, nearing the Pacific Ocean, the Columbia River carves a long, deep canyon through the heart of the Cascade Range. Here the river flows between dramatic, high walls of basalt laced by ribbons of delicate waterfalls on the south side of the river and low hills fractured by areas of extensive landslides on the north. The Columbia River Gorge was created when the erosive action of the Columbia was able to outpace the uplifting of the young Cascade Range and maintain a channel linking waters of central Washington's Columbia Plateau to the Pacific Ocean.

The oldest rocks exposed in the gorge are those that, starting about 38 million years ago, were deposited by eruptions of early Cascade volcanoes. The area was a vast lowland at the time, crossed by meandering streams that originated in the eastern part of the state and flowed west to the coastline near what is now the location of Highway I-5. Lava and mud flows spread across this flat terrain as wide rings around volcano vents. They built layer upon layer of sedimentary rocks composed of small boulders and cobbles encapsulated in a slurry of pumice fragments and volcanic tuff. Although the thickness of these flows increased rapidly, accumulating to

A trail with fifty-three switchbacks leads to the top of Beacon Rock.

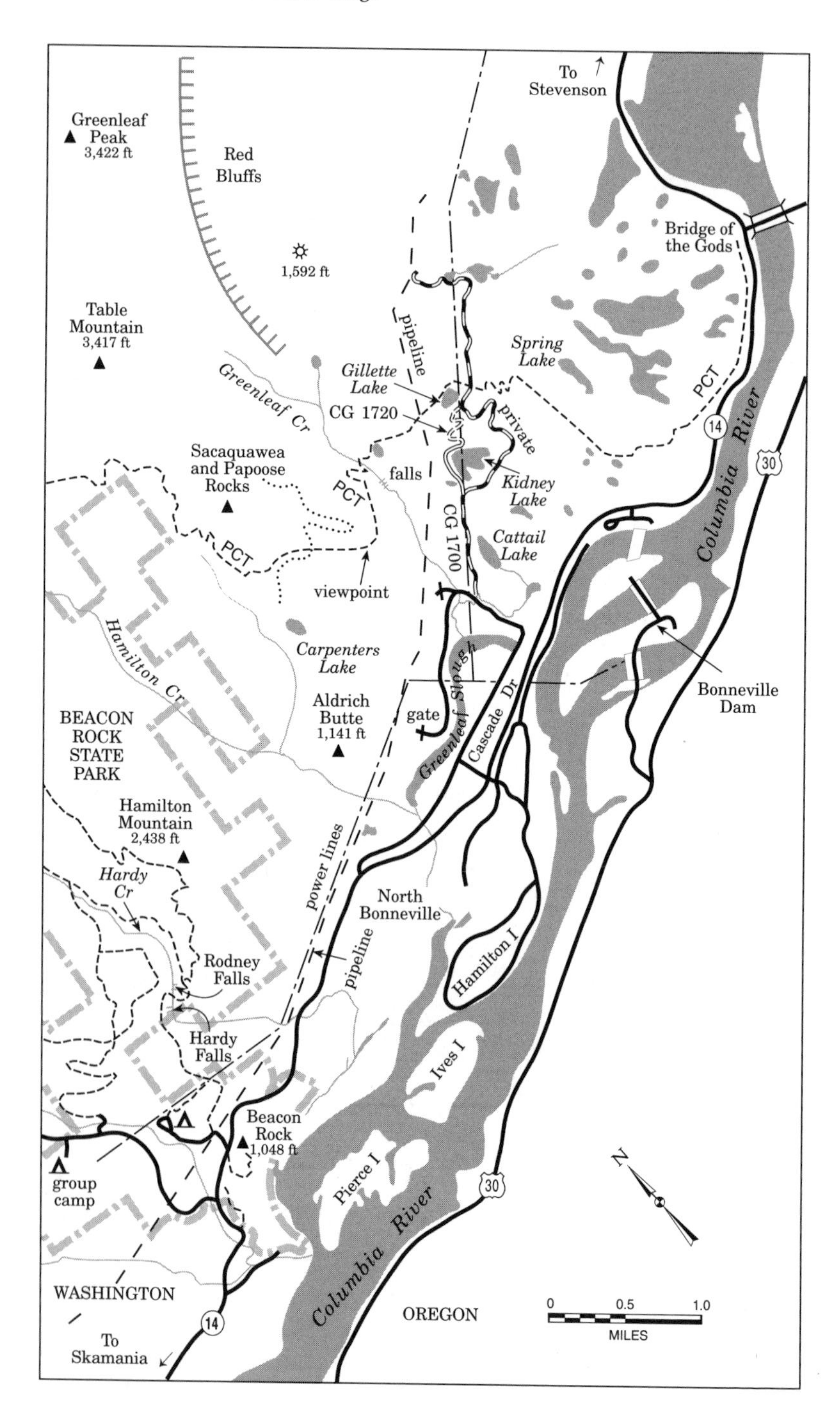
To Stevenson
Greenleaf Peak 3,422 ft
Red Bluffs
1,592 ft
Bridge of the Gods
Table Mountain 3,417 ft
pipeline
Spring Lake
Gillette Lake
Greenleaf Cr
CG 1720
private
PCT
14
Columbia River
30
Sacaquawea and Papoose Rocks
falls
Kidney Lake
PCT
CG 1700
Cattail Lake
PCT
viewpoint
Hamilton Cr
Carpenters Lake
Bonneville Dam
Aldrich Butte 1,141 ft
gate
Greenleaf Slough
Cascade Dr
BEACON ROCK STATE PARK
Hamilton Mountain 2,438 ft
power lines
Hardy Cr
North Bonneville
pipeline
Hamilton I
Rodney Falls
Hardy Falls
Ives I
Beacon Rock 1,048 ft
group camp
Pierce I
30
N
Columbia River
WASHINGTON
OREGON
0 0.5 1.0
MILES
14
To Skamania

depths of 800 to 1,300 feet, the elevation gain was slight, as the underlying crust sank at about the same rate.

About 17 million years ago volcanic activity waned and a gradual uplift of the Cascades began, possibly triggered by the same upwelling of magma that created the Columbia River flood basalts. These deep floods of basalt spread across south-central Washington and northern Oregon about 16 million years ago, reaching the coast and invading the soft marine muds. Although most of these basalts were emplaced just above sea level, subsequent folding and uplift of the South Cascades has raised these strata to between 2,000 and 5,000 feet of elevation. This same uplift tilted rock layers in the Columbia Gorge region, angling them downward from north to south. As a result, when the Columbia River cut through this rock, steep cliffs resulted on the south side of the river, while the downsloping layers to the north were subject to huge landslides.

Beacon Rock

Hiking trail to the top of a volcanic plug

Rating: (M); hikers
Distance: 1 mile
Elevation: Trailhead 200 feet, summit 1,048 feet
Maps: USGS Beacon Rock, USFS Wind River Ranger District
Driving Directions: Drive Highway 14 29.9 miles east from US 205 to Beacon Rock State Park.

This impressive basalt monolith on the north shore of the Columbia River was formed about 1.6 million years ago, when magma pressed upward through relatively soft older rock strata, feeding lava flows from a volcano vent at the site. The magma conduit eventually solidified into a mass of platy, blocky, columnar-jointed olivine basalt. Subsequent mud and lava flows overlaid the area with softer, more easily eroded strata. Over time, the Columbia River carved through softer layers and washed them away, exposing the hard old volcanic plug.

The landmark was called Che-che-op-tin by local Native Americans and was incorporated into their legends long before it was seen by whites. As the story goes, the rock was first climbed by a Native American princess, Wahatoplitan, to save her son from her angry father, who disapproved of her marriage. Both Wahatoplitan and her son died atop the rock. The sound of wailing, mourning their deaths, is still heard when Chinook winds from the east whip over the summit.

The Lewis and Clark Corps of Discovery, the first American party to enter the country by land, noted this distinctive landmark in their journals on October 31, 1805, first calling it Beaten (then later correcting it to Beacon) Rock. In 1811, the John Jacob Astor expedition renamed the rock for Inshoack Castle, and it was known as Castle Rock until the original name was restored in 1916.

To crass commercial interests, including railroad barons, whose tracks ran along the rock's base, the rock was just a huge, potential quarry for railroad ballast and highway beds. Although it was saved from this fate by its purchase by Henry Biddle in 1915, as late as 1931 the Corps of Engineers eyed it as a source of rock for constructing jetties at the mouth of the Columbia.

When Biddle bought the rock, he envisioned building a trail to the top. After investing $10,000 and two years' time in the project, he completed the 0.8-mile trail up the south face in 1918. Although many people wanted the rock to be designated a state park, the concept was thwarted by one of Washington's anti-park governors, Ronald Hartley. Heirs of Biddle deeded the property to the state of Oregon in 1932 (even though it was located in Washington) to assure its future protection. With the election of a more enlightened administration in Washington state, the property became a state park in 1935.

Today, Biddle's fifty-three switchbacks, some overlapping one another, lead thousands of hikers annually to the top of the rock to see the

Bridge of the Gods Legend

Many variations of a Native American legend describe geological features in the southwest Cascades. One is as follows: The Great Spirit had two sons, Wy-east (Mount Hood) and Pah-toe (Mount Adams). He shot two arrows into the air and sent his sons to occupy lands where the arrows landed, Wy-east to the south of the Columbia River, and Pah-toe to the north side. For many years the brothers lived in peace in their respective domains, and the people, plants, and animals surrounding them prospered.

Enter the ingenue, an attractive miss who came to a valley between the two brothers. Both were smitten with love and began to fight over her. They stomped the ground, shaking it violently (earthquakes of unknown Richter levels!), and threw rocks and fire at each other. When the altercations were over, all the animals, plants, and people that had lived happily on their slopes saw the destruction wrought by the fight. As a capper to the melee, the stomping and shaking of the earth triggered an avalanche that dammed the river between the two. The river tunneled through the earthen dam, forming a natural bridge joining both sides of the river.

The Great Spirit became aware of this sorry state of affairs, banished the young woman to a cave, and decreed that the bridge between the brothers' domains would remain as a symbol of peace between them and serve as a means for the people from the two sides to cross and mingle. However, if the brothers could not maintain a civil relationship, the bridge would be destroyed! Finally, as a reminder of the foolishness of pursuing youthful beauty, he put an old crone, impressive views of the Columbia River. The sheer cliffs on the northeast and southeast sides of the rock remain one of the most difficult rock-climbing challenges in southwest Washington.

Hamilton Mountain

Hiking trail past waterfalls to views of volcanic strata

Trail: Hamilton Mountain Trail
Rating: (D); hikers
Distance: 3.4 miles
Elevation: Trailhead 600 feet, Hardy Creek 1,000 feet, Hamilton Mountain 2,438 feet
Maps: USGS Beacon Rock, USFS Wind River Ranger District
Driving Directions: Follow directions to Beacon Rock, preceding.

Hike part way up the trail to Hardy Creek, where spur trails lead uphill to enchanting Rodney Falls and downhill to equally beautiful Hardy Falls,

Loo-wit—the keeper of the fire—by the bridge to guard it.

Well, love was not to be thwarted, and Wy-east found a way to reach the beautiful maiden's cave and engage in trysts with her. The Great Spirit discovered this intrigue, and tacitly permitted the relationship to continue. One day, however, Pah-toe found out about these goings on, and the brothers tossed off their white glacial cloaks and went at it again with fire and brimstone. Their fight shook the ground so hard that the bridge across the river collapsed and forced its flow into a narrow channel, marked by the Narrows and the once-dreaded Cascades of the Columbia, after which the Cascade Range was named. (This area has since been covered by the reservoir behind Bonneville Dam.)

The Great Spirit was, in modern terms, ticked off and transformed the object of the brothers' affections into a mountain near Pah-toe, now known as Sleeping Beauty. Loo-wit, the old woman who had tried valiantly to protect the bridge, was given her wish to become young and beautiful and was moved west to become the lovely, albeit restless, cone of Mount St. Helens.

No doubt there is some geological basis to the Bridge of the Gods legend. Adams and Hood certainly went through eruptive phases that would have devastated plant, animal, and human life on their slopes. Earthquakes did, in fact, trigger major landslides on the north side of the Columbia that diverted its flow and for a time dammed the river. There is no evidence that the Bridge of the Gods actually existed, but it is within the realm of possibility. The beauty, and more recently the restlessness, of Mount St. Helens can certainly be attested to.

Hikers are dwarfed by the immense cliffs of Hamilton Mountain.

or continue an arduous climb to the top of Hamilton Mountain and views from atop its 1,000-foot-high vertical southeast face to the Columbia River and Bonneville Dam.

Begin the hike from the trailhead at the lower picnic area on the east side of the campground road or from the campground loop itself. Both routes merge in 0.2 mile. The path climbs steadily north for another 0.7 mile to reach a bridge over Hardy Creek. A short boot path 100 yards west of the bridge leads downhill to a viewpoint of the creek as it spills over a rock lip, creating the slender, 80-foot-high ribbon of Hardy Falls. Just above the bridge are the series of cascades that make up Rodney Falls.

Now the serious hiking begins. Nearly three-dozen switchbacks are needed in the next mile to gain the top of the south shoulder of Hamilton Mountain. As the trail continues north on this ridge top it skirts the edge of the 650-foot vertical east face. Wide views show the continuation of this sheer cliff to the north, below the summit of the mountain.

This face displays an excellent cross-section of local geology. The wall is primarily made up of sedimentary rock formed between 22 and 17 million years ago. At that time long periods of heavy precipitation triggered floods that covered the area with thick layers of mud flows composed of fine volcanic particles, cobble-sized fragments of pumice, and ash washed from the slopes of older volcanoes of the Cascades to the northwest. Near the summit of Hamilton Mountain, and along the ridgeline to the north, this sedimentary rock is overlaid by flows of Yakima Basalt. This is the northwest edge, within the Columbia River Gorge, of the massive basalt flows that flooded the Columbia River basin in central and eastern Washington and Oregon some 16 million years ago.

In another 0.2 mile the route starts its final ascent to the top of the mountain; the steep climb requires another two-dozen switchbacks before reaching the 2,438-foot-high summit of Hamilton Mountain. Here are unsurpassed views of the Columbia River Gorge, Beacon Rock, Bonneville Dam, and the massive landslide slopes south of Table Mountain and Red Bluffs that briefly dammed the Columbia River long ago.

Table Mountain, Red Bluffs, and Greenleaf Mountain

Hiking trail across a massive landslide

Trail: PCT #2000
Rating: (M); hikers, saddle and pack stock
Distance: From road CG 1700 to #2000 0.5 mile, to the west through areas of geological and scenic interest 2.5 miles
Elevation: Kidney Lake 260 feet, viewpoint 640 feet
Maps: USGS Bonneville Dam, USFS Wind River Ranger District
Driving Directions: Take Highway 14 east from US 205 33.8 miles to North Bonneville. Turn north on Dam Access Road, cross under the railroad tracks, and in 0.2 mile turn east on Cascade Drive. Follow this road east, then north for 0.6 mile, where CG 1700 (1½-lane gravel) heads northeast under the Bonneville power lines. In another 0.6 mile, the road is signed "Private, No Trespassing." About 50 yards before the sign, road CG 1720 (1-lane dirt) continues northeast under the power lines. Vehicles with reasonable clearance can follow it for another 0.2 mile to Kidney Lake, where it becomes impassable.

Drivers whizzing by on Highway 14 don't have time for a good look at the countryside north of Bonneville Dam. This reasonably easy hike, just a short distance off the highway, provides an opportunity to see some of the major geological elements that shaped this portion of the Columbia River Gorge and to enjoy some scenic views of the river. Turn north onto Dam Access Road at North Bonneville, and just after crossing under the railroad tracks go east on Cascade Drive. In a few hundred yards pull off the road near the head of Greenleaf Slough and look uphill to the north (binoculars will help immensely). The first point of geological interest can be seen here.

The highest point on the horizon is the triangular summit of Table Mountain, where a sheer 600-foot-high face on the southeast side of the peak exposes at least six different layers recording the geological history of the region. The sandstone layers low on the face were emplaced between 22 and 17 million years ago by mud flows carrying debris from the flanks of older volcanoes to the northwest. The multiple strata higher up

Rocky pinnacles on a ridge of Table Mountain resemble Sacaquawea holding her papoose.

in the face are layers of basalt from the Yakima flows, the western finger of the immense flood of basalt lavas that covered the central and eastern portions of Washington between 17 and 15 million years ago.

At the lower end of the ridge south of the Table Mountain summit is a distinctive finger of rock rising more than 100 feet above the surrounding landscape. Atop this finger are two pinnacles, named Sacaquawea and Papoose Rocks, honoring the Shoshone woman who guided Lewis and Clark. Farther north along the ridge is another rock fin, part of the same formation, a basalt dike that intruded upward into the native rock about 20 million years ago.

Continue on to the road end at the north end of Kidney Lake and walk uphill on a steep, rough jeep road to the northeast. In 0.5 mile the route rejoins private road CG 1700, which accesses a large gravel pit a mile uphill. Hike the edge of the road uphill for 0.2 mile to where it crosses the PCT, then head west on the trail toward Gillette Lake. Sacaquawea and Papoose Rocks are framed against the skyline just after crossing under power lines north of the lake.

Uphill, north of the lake, the slope is a huge field of basalt blocks and talus, covered with moss and shrubs. This is the surface of a massive landslide that broke from the face of Table Mountain and Red Bluffs about 1,100 years ago. The landslide possibly was triggered by a major earthquake along the junction of the North American and Juan de Fuca plates that also impacted several other spots in the Pacific Northwest about this same time. The quake caused large sections of Table Mountain and Red Bluffs to slip to the southeast, temporarily blocking the flow of the Columbia River and probably giving rise to the natives' Bridge of the Gods legend.

The trail continues a gradual ascent to the west, crosses a natural gas pipeline right-of-way, then weaves above the shore of an unnamed lake, all the while picking its way through old overgrown landslide debris. After crossing a sturdy bridge across Greenleaf Creek, the way snakes uphill to an open point with an excellent overview of the old landslide path drifting down to the present site of Bonneville Dam. From here the PCT continues through dense forest that blocks scenic views. This is a good spot to turn around.

4

The Wind River Country

From its headwaters west of the Indian Heaven Wilderness, the Wind River flows some 25 miles, first south, then southeast to join the mighty Columbia near Carson, Washington. It is a major tributary of the Columbia River on the west side of the Cascade crest.

Beginning roughly 38 million years ago, an explosive volcano in the vicinity of the South Cascades spewed out vast amounts of andesite and dacite volcanic debris and ash flows. Stream currents sorted and eroded this material and, as it accumulated over the next 20 million years, it became layered sedimentary rock. These layers built up to more than 10,000 feet in thickness in the subsiding trough of the early Cascade Range. Streams from the north and northwest washed successive layers of andesite sands and conglomerates over the surface, primarily in the south end of the Wind River area. Subsequent erosion shaped the landscape into low rolling hills.

A massive flood of basalt lavas flowing west across Washington and Oregon, starting about 17 million years ago, reached as far west as the seacoast from Hoquiam, Washington, to Newport, Oregon. Fingers of these lava flows swept over the eroded surface of the Wind River region, accumulating to depths of over 1,000 feet, predominately in canyons in the southern portion of the area. This was followed by uplifting of the range. The pressures associated with this uplifting occasionally triggered faults and folding.

Deep-seated magma, formed from a subducting oceanic plate, rose beneath the area, pressing upward through faults and weak spots in the overlaying strata, then congealing below the surface to form domes, dikes, and sills. These intrusions started as early as 25 million years ago, but most were emplaced between 5 and 3 million years ago. Yet another period of folding occurred between 2 million and 500,000 years ago. This was followed by a new period of volcanic eruptions, mainly of basalts, from the shield volcanoes that form many of today's major recognizable terrain features. Starting about 340,000 years ago, these shield lavas were in turn overlaid with olivine basalt from younger vents, many located along the lines of deep-seated faults in older rock. A number of these vents capped their eruptive cycle with the cinder cones that ejected localized blocky lava, clinkery fragments, and tephra, building up some of the traditional, volcano-shaped peaks in the area.

Periods of intermittent, widespread glaciation that began about 200,000 years further altered the shape of the previous volcanic activity. The Wind River area saw its most recent eruptions within the last 10,000 years when young vents, such as West Crater, ejected localized flows of blocky basalt that are still visible today, having not yet been totally overgrown by vegetation.

The geological features described in this chapter lie west of the Wind River drainage. The easiest approach to the area is via Wind River Highway, which leaves Highway 14 at Carson, 50 miles east of Vancouver. It follows the Wind River drainage upstream to Government Mineral Springs, where it continues north as FR 30. From the north, the shortest approach

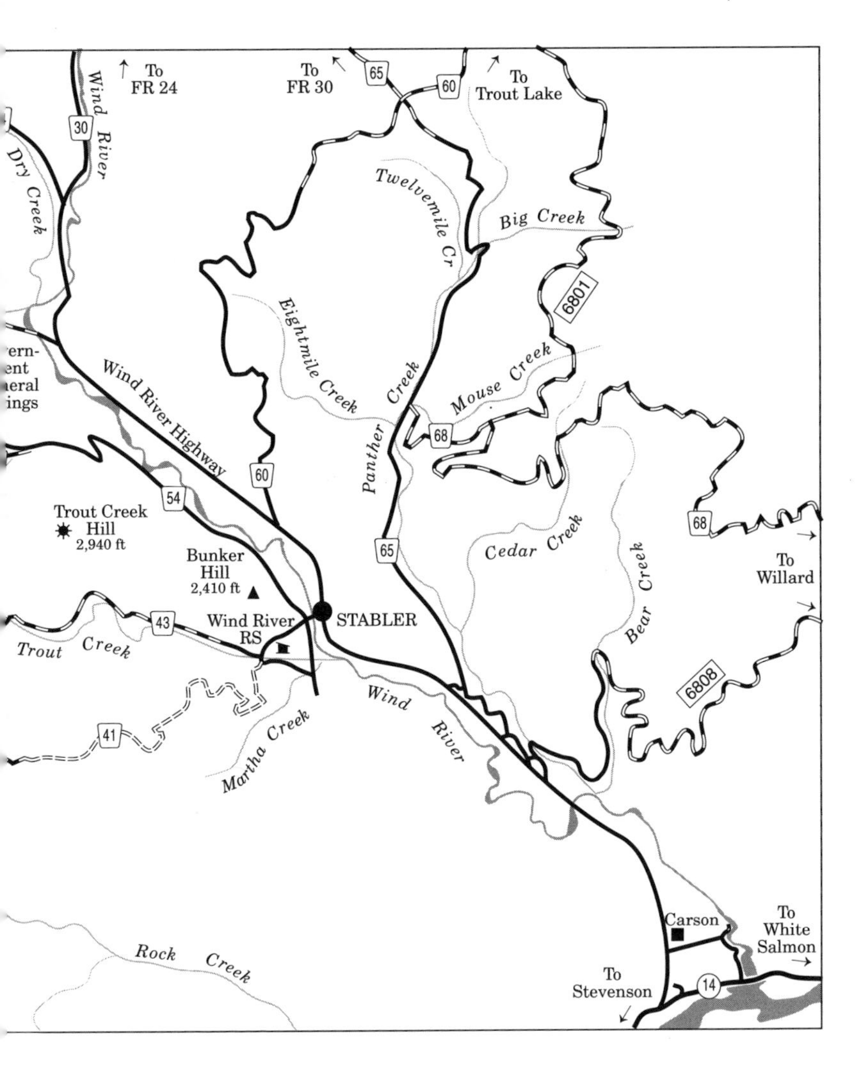

is via FR 51 from Curley Creek, which leaves Lewis River Road, FR 90, 36.3 miles east of Cougar, then climbs uphill 7.5 miles to join FR 30, 15.8 miles north of Government Mineral Springs. All these roads are paved.

A more challenging route from the west leaves Highway 503 as Healy Road at Chelatchie, 30.5 miles east of Woodland, then proceeds east to the forest boundary, where it becomes FR 54. After 14.4 miles, some paved, but most single-lane gravel, the junction with FR 58 is reached. Both routes continue east, at different altitudes, to rejoin in another 11 miles at the junction of FR 54 and FR 5407. FR 54 then snakes downhill to the east for 13.5 miles to Wind River Highway.

Wind River Hot Springs

Scramble to a hot springs

Rating: (M); hikers
Distance: 0.6 mile
Elevation: 270 feet
Maps: USGS Carson, USFS Wind River Ranger District
Driving Directions: Follow Highway 14 east from Stevenson, passing the two exits to Carson, the most easterly of which is Hotsprings Avenue. In another 0.5 mile, turn north on Berge Road (2-lane paved), follow it uphill for 0.8 mile, then turn west onto Indian Cabin Road (1-lane paved). The road runs under the Bonneville power lines for 0.5 mile, then twists downhill, deteriorates to gravel, and arrives at a T-intersection in another 0.4 mile. Go right, cross a bridge over the Little Wind River, and in a few hundred feet reach a large dirt parking lot.

No posh resort this! On the contrary, river bank rocks here have formed basins that collect hot spring flows seeping from the hillside at spots along the river. There are a few small bowls, barely large enough to warm your feet, and one large spa-sized grotto that can accommodate several soaking bodies. The remoteness of the pool and the effort required to get to it make skinny-dipping both acceptable and popular. If it offends you, don't go.

Hot springs found on both sides of the river in this area originate from underground springs heated in the contact zone between a thick, 30-million-year-old base of andesite flows and volcanic debris that underlies the region and a much younger quartz diorite magma plug that intruded into the older rock about 5 million years ago. The west bank of

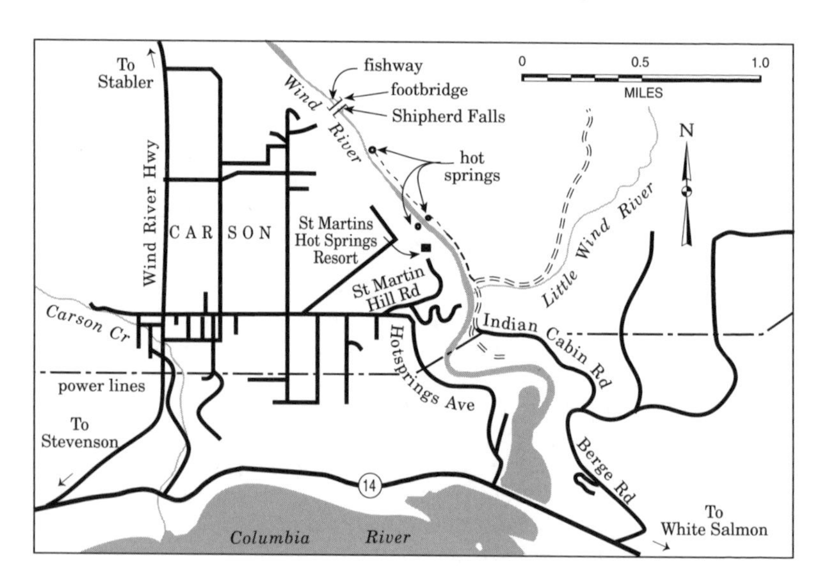

Water in Wind River Hot Springs is naturally heated deep in the earth. Enjoy!

the river consists of the 350,000-year-old lava flow from the Trout Creek Hill volcano, which filled the original Wind River valley.

Trailhead parking lot amenities consist of a decrepit outhouse, a signboard, and a collection box. Ignore the collection box at your peril. This is private property, and failure to deposit an easement fee is regarded as trespassing; your vehicle will get ticketed or towed. As of 1994 the fees were $2 per vehicle plus $2 per person (including the driver). The trail is unmaintained, and there is poison oak en route. Sandals are discouraged, with good reason—scrambling over the rough, slippery river rock is far safer and more comfortable in sturdy boots.

The narrow, dirt path traverses an ever-steepening hillside above the river, and the farther you go the more difficult the path becomes. Spurs periodically drop down to the river's edge. Take one of the earliest you find and slide down to the rocky bank; boulder-hopping alongside the river is safer and somewhat easier. Make sure you can recognize which of the several spurs to take on your return, because there are no trail markings. The route is more difficult—or even impossible—during periods of high water.

After about ½ mile of rock scrambles the largest of the hot spring pools is reached. Upstream you can see a weir with a suspension bridge above; both are part of a local fish hatchery. Slip into your preferred bathing attire or non-attire, step into the 120-degrees-Fahrenheit pool, relax, and enjoy.

Government Mineral Springs

Short footpaths to mineral springs

Rating: (E); hikers
Elevation: 1,300 feet
Maps: USGS Termination Point, USFS Wind River Ranger District
Driving Directions: Take the Carson exit from Highway 14, and drive northwest on Wind River Highway (2-lane paved) for 14.4 miles to the intersection with Mineral Springs Road (2-lane paved). Continue northwest for 1.1 miles to Government Mineral Springs.

A glamorous destination resort in the early 1900s is now a picnic area and de facto campground in an open field. At the junction of Wind River Highway and Mineral Springs Road, continue north a mile on Mineral Springs Road to a one-way loop through the Government Mineral Springs recreation area. After a ½-mile passage through magnificent old-growth Douglas-fir and western red cedar, a spur road leads right to the Iron Mike well and a path to Little Iron Mike and Bubbling Mike spring.

Little Iron Mike mineral springs is surrounded by moss-covered rocks.

The old Iron Mike well is northwest of the parking area at the end of the spur road, under the cover of a rustic, log gazebo. You can still pump up the smelly, sulfur-laden water that was acclaimed for its curative powers during the resort's heyday. Cure you? Questionable. Kill you? Taste it and you'll think it would!

A short trail heads east. Little Iron Mike, which is signed, is colorful, although not impressive. A slow flow of bubbles filters up through a thick layer of orange algae at the bottom of a creeklet sprinkled by a color-contrasting collection of brilliant green, moss-covered rocks. Bubbling Mike, which is unmarked, is a few hundred feet downstream on the south edge of Trapper Creek. The distinction between the two springs is primarily the volume of bubbles rippling up from the orange-colored creek bottom. The source of the rather tepid mineral springs lies in the contact of clay

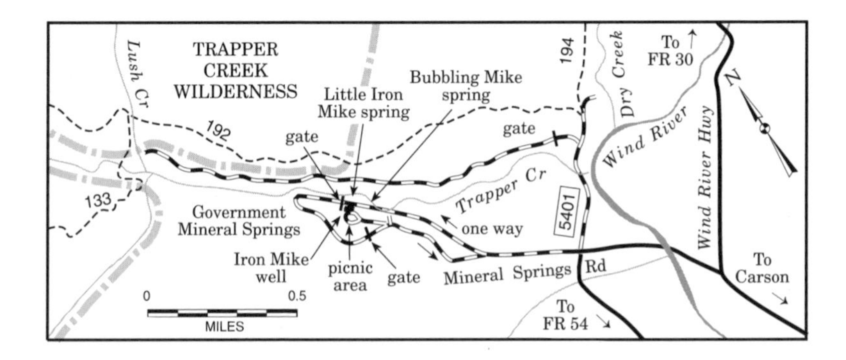

and gravel deposits with underlying ancient volcanic debris and lava flows. Other contacts with lava sources below the nearby Trout Creek Hill and Soda Peaks volcano vents may contribute to the formation of the springs.

Return to the one-way loop and drive a short distance, taking the next right to its end in an open field. This was the site of the Government Mineral Springs Resort, which included a three-story hotel with a dance pavilion, mineral baths, riding stables, a formal garden, and an adjoining campground. Guests arrived in horse-drawn wagons from Carson, 15 miles to the south. Rumor had it that during Prohibition the resort was a secret haven for boozing and gambling! All of this ended in 1935, when the resort burned to the ground in a tragic fire. The picnic area and campground field and the Iron Mike gazebo are all that remain today.

Roads upstream on either side of Trapper Creek are gated; summer homes that line both sides of the creek are private inholdings that predate the creation of the adjoining Trapper Creek Wilderness.

Bunker Hill

Hiking trail to the top of a volcanic plug

Trails: PCT #2000, Bunker Hill Trail #145
Rating: #2000 (M); #145 (D); hikers, saddle and pack stock
Distance: 1.7 miles
Elevation: Trailhead 1,140 feet, summit 2,410 feet
Maps: USGS Stabler, USFS Wind River Ranger District
Driving Directions: Take Wind River Highway northwest from Carson for 8 miles to Stabler, then turn west on Hemlock Road (both 2-lane paved) and in 1.1 miles reach the Wind River Ranger Station. Just beyond the ranger station head northwest on FR 43 to FR 4300417 (both 1-lane gravel), which soon intersects Trail #2000.

A grinding uphill drag that gains almost 1,200 feet in a little over 1 mile offers outstanding views of the Wind River valley and major volcano vents and flows to the northwest. From the point that FR 4300417 crosses the

Bunker Hill, which rises above the surrounding Wind River countryside, is an igneous plug. Its summit offers grand views of surrounding volcanic features.

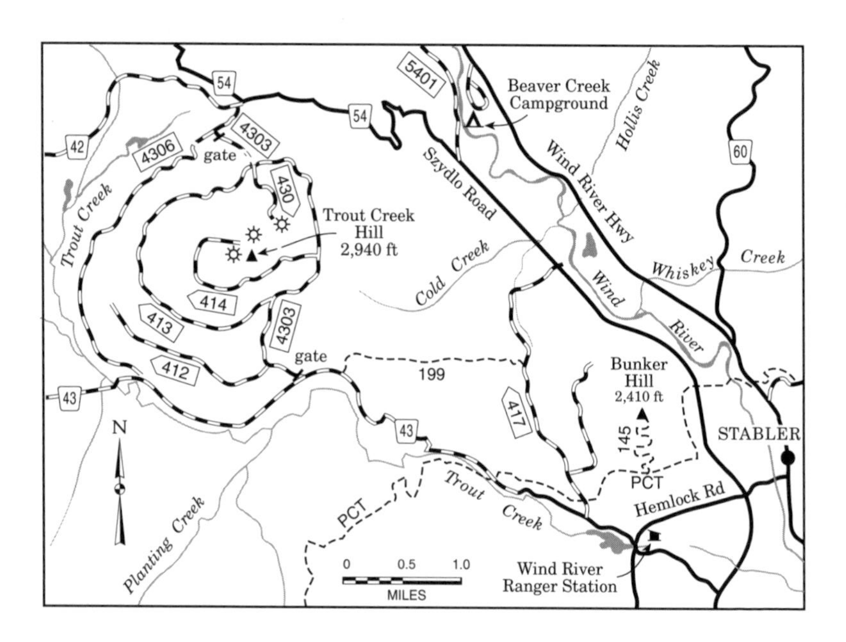

PCT, take the latter along the edge of a logged patch into second-growth forest. In 0.5 mile turn uphill to the north on Trail #145. The grade starts steep, and gets steeper, as long switchbacks swing up the near-vertical south nose of the mountain. Periodic breaks in the tree cover offer wide, impressive views to the south across tree farms of the Wind River Ranger District and down the lower river drainage.

As the route snakes upward toward the short summit ridge, it crosses a 400-foot-high cliff where any misstep would be trouble. A fire lookout cabin, built around 1940, stood atop the summit until it was abandoned in the 1950s.

The hill itself is an igneous plug pushed up between 25 and 20 million years ago through older layers of lava flows and volcanic debris. It stands as an island amid much younger basalt flows from the Trout Creek Hill shield volcano that flooded the Wind River valley as far south as the Columbia River about 136,000 years ago.

Trout Creek Hill

Road walk and cross-country route to the top of a shield volcano

Rating: (M); hikers, mountain bicycles, saddle and pack stock
Distance: 2.6 miles
Elevation: Wind River Ranger Station 1,110 feet, top of Trout Creek Hill 2,940 feet
Maps: USGS Lookout Mountain, Stabler; USFS Wind River Ranger District
Driving Directions: Follow the directions in Bunker Hill, preceding, to reach the Wind River Ranger Station. From the ranger station continue northwest on FR 43 (1-lane gravel) for 3.2 miles to FR 4303. FR 4303 is gated and closed to motorized vehicles at this point.

The unique, roughly circular shape and lack of high ridges connecting it to surrounding terrain confirm that Trout Creek Hill is a free-standing shield volcano. But not just *any* shield volcano—this one had a major impact on the lower Wind River drainage.

About 350,000 years ago, coarse olivine basalt started erupting from this site. This lava flood spread southeast, split around the older intrusion of Bunker Hill, then rejoined and continued its course downstream along the bottom of the Wind River drainage, filling it to a depth of nearly 300 feet. The broad lava stream continued south, past the present location of Carson and into the Columbia River Gorge, where it formed a dam that temporarily blocked the flow of the Columbia River. Backed-up river water created a lake, and sand and gravel carried by the Wind River created an alluvial fan over 100 feet thick near its mouth. Finally, the Columbia breached the dam and drained the lake.

A logging road leads nearly to the top of the hill. From the gate at the junction of FR 43 and FR 4303, hike FR 4303 north for 1 mile, then head west on FR 4303414, which circles around the south and west sides of the hill for another 1.6 miles before ending about 150 feet below its top. As the roads wind around the mountain, views from clearcuts help establish its place in the geology of the Wind River area.

A short cross-country hike from the road end leads to the summit of Trout Creek Hill. The soil changes subtly as you approach the summit. Traces of tephra clinkers mark a transition in eruptive phases, culminating in the growth of two small cinder cones atop the summit of the older shield. Be extremely careful exploring the summit; a 60-foot-deep pit with near-vertical walls marks one of the volcano vents.

West Crater

Cross-country route to a glacial cirque and a volcano crater

Rating: (D); hikers
Distance: To crater rim 0.2 mile
Elevation: FR 34 saddle 3,840 feet, rim of crater 4,160 feet, center of crater 4,100 feet
Maps: USGS Bare Mountain, Lookout Mountain, Stabler; USGS Wind River Ranger District
Driving Directions: Drive Wind River Highway north from Carson for 7.5 miles to Stabler, then turn west on Hemlock Road. In 0.2 mile, turn northwest on FR 54 (Szydlo Road) (1-lane paved with turnouts), and continue northwest for 12.9 miles to FR 34 (1-lane gravel). Here turn south, and in 1.8 miles pull off into a saddle on the west side of West Crater.

Here is a young (8,000-year-old) volcano vent wrapped around a small, sunken crater. A visit to the crater floor creates the spooky feeling that tiptoeing may be appropriate to avoid waking the (hopefully) dormant fires beneath. As FR 54 climbs steep slopes enclosing deep canyons west of the Wind River, the hilltop at the head of the Trout Creek drainage takes on the distinctive truncated-cone shape of a one-time volcano. Suspicions are further confirmed by a glance down to the headwaters of Trout Creek to a tongue of dark, blocky basalt, not yet hidden by encroaching forest, dropping down the hillside.

As FR 54 reaches a saddle 15 miles from Wind River Highway, FR 34 branches south. At a roadcut in a few hundred yards note the blocky andesite strata that are part of the 30- to 15-million-year-old base structure underlying most of this portion of the South Cascades. FR 34 crosses a saddle between Hackamore and Trout creeks, then winds around the clearcut west side of West Crater, slicing across a loose, extremely steep slope that drops abruptly for more than 300 feet to the Hackamore Creek drainage. Some drivers may feel uncomfortable on this 0.2-mile stretch of road.

After crossing this slope, park on the east side of the road in a small flat. Head cross-country to the east through huckleberries and moderately dense fir. In about 300 yards break into an open meadow, some 200 yards in diameter, surrounded by a 100- to 250-foot-high forested rim. The outer rim of the meadow is a lush growth of beargrass that wraps around a smaller inner meadow in a 3-foot-deep depression that is covered with grass, a few hummocks of beargrass, and a smattering of lupine and harebells. Although this has the appearance of a crater, in reality it is an old glacial cirque from which ice once flowed northwest into Hackamore Creek; most of the lip of the cirque was later blocked by the younger volcano.

On the northeast rim of this cirque is West Crater, the source of the basalt flows you saw earlier from FR 54. To reach the crater of this volcano, return to the parking spot, and head cross-country to the east, climbing steeply uphill along the near edge of the clearcut. The ground cover of knee-deep huckleberry bushes and young, 3-foot-high (and growing) fir is not easy going, but the views en route and the peek into the crater are worth the effort.

The deep, wide Canyon Creek drainage, which drops more than 2,500 feet to the northwest, is a crazy-quilt of deep green old-growth forest, paler replanted clearcuts, and bald, freshly logged slopes. Logging atop the ridge along the north side of the drainage has exposed the cone-shaped summits of West Point, and Point 3685, south of Calamity Peak; they almost

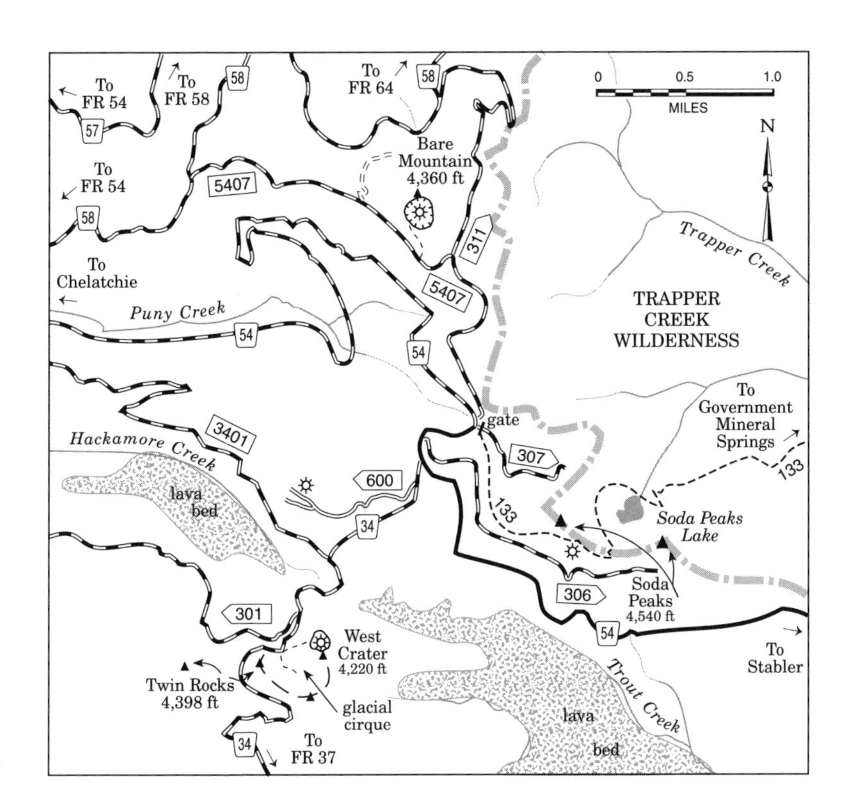

The sunken center of West Crater holds a small meadow, seen here from the edge of the crater.

shiver in their nakedness. To the north, the truncated summit of St. Helens dominates the horizon. In about 100 vertical feet the 100-yard-wide crater rim is reached. Steep, partially wooded walls drop some 50 feet to a tiny meadow at the heart of the dormant volcano. A few outcrops of basalt serve as a reminder of the crater's violent past.

Soda Peaks

Hiking trail past a volcano vent to a lake in a glacial cirque

Trail: Soda Peaks Lake #133
Rating: (D); hikers
Distance: From FR 54 to Soda Peaks 1.2 miles, to Soda Peaks Lake another 0.8 mile; trail continues east for 4 miles to Government Mineral Springs
Elevation: FR 54 trailhead 3,680 feet, Soda Peaks 4,540 feet, Soda Peaks Lake 3,760 feet
Maps: USGS Bare Mountain, Termination Point; USFS Wind River Ranger District
Driving Directions: Follow directions to West Crater, preceding, as far as the junction of FR 54 and FR 34. Continue northwest on FR 54 for another 0.4 mile to the trailhead at the junction of FR 54, FR 5407, and FR 5400307.

A newly reconstructed trail climbs into the Trapper Creek Wilderness to a saddle next to a volcano vent then drops to an alpine lake filling a glacier-carved cirque, before descending to the main entry-point into the wilderness. The junction of FR 54, FR 5407, and FR 5400307 is the start point of Trail #133; fill out your wilderness permit here.

The path wanders through a logged-off flat then makes a gradual ascent east to trace the uphill side of yet another clearcut, which in late summer is brush-stroked a brilliant purple by massive fields of lupine blossoms. The route soon climbs rapidly to a narrow saddle between the two major summits of Soda Peaks. Ridge-top views extend from Rainier on the north to Adams on the east.

The timbered summit to the south side of the saddle is an old

Lupine

volcano vent that erupted a flow of olivine basalt, primarily to the east and southeast, some 360,000 years ago. From the saddle, Trail #133 switchbacks down steep talus slopes to the glacial cirque wrapped around Soda Peaks Lake; the azure surface reflects the surrounding peaks.

From the lake the trail continues east to join Trail #192 at Trapper Creek. From there the route descends a short distance to the trailhead near Government Mineral Springs.

Bare Mountain

Cross-country route to a volcano crater

Rating: (D); hikers
Distance: 0.2 mile
Elevation: FR 5407 trailhead 3,940 feet, crater rim 4,200 feet
Maps: USGS Bare Mountain, USFS Wind River Ranger District
Driving Directions: Follow directions to West Crater, earlier, as far as the junction of FR 54 and FR 34. Continue northwest on FR 54 for another 0.4 mile to the junction of FR 54, FR 5407, and FR 5400307. Take FR 5407 (1-lane gravel) northwest from FR 54, pass FR 5800311, and park in 1.4 miles, just before the road crosses the bare southwest face of Bare Mountain.

A short cross-country trek reaches the rim of a dormant volcano, with views down into one of the largest young craters in this part of the Wind River country. FR 5407 skirts the southwest side of the mountain, but the bare slopes immediately uphill are brutally steep. The most reasonable approaches are from either side of the steep southwest face. The approach described here is from its south edge.

Start the cross-country trek from the roadside parking area. Head diagonally uphill to the northwest, taking advantage of short segments of animal trails, while plowing through knee-deep huckleberry bushes and 10-foot-high second-growth. Stay the course, and eventually arrive at the rim of the crater, somewhere near its low point on the southwest lip.

Look down into one of the largest and best-formed of the younger craters in this part of the forest. Some 120 feet below, at the bottom of wooded walls, is a circular open meadow occupying the heart of the former volcanic cauldron. On the opposite rim are knobby basalt outcrops, remnants of the lava that once pushed out from the volcano core. The basalt flows from this vent, which erupted about 8,000 years ago, were directed mostly to the north into the headwaters of Siouxon Creek, where they continued downstream for nearly 2.5 miles before banking against sidehills in the Siouxon and Calamity creek drainages.

For the adventuresome, the loop trip through open forest around the top of the crater rim is just over ½ mile and can include a scramble to the mountain's highest point, the rock knob on the northwest side.

West Point

Hike on an abandoned trail past dikes to an intrusive dome

Rating: (M); hikers
Distance: From FR 5704315 to the summit of West Point 1 mile
Elevation: End of FR 5704315 3,420 feet, summit 3,850 feet
Maps: USGS Siouxon Peak, USFS Wind River Ranger District
Driving Directions: From the east, follow directions to Bare Mountain, preceding. Continue west on FR 5407 for another 1.6 miles to FR 58 (both 1-lane gravel). In another 1.8 miles, turn west on FR 5704 (1-lane gravel), and follow it west 2 miles to FR 5704315 (1-lane dirt). Continue northwest on this narrow, rough track to a roadblock in 0.2 mile.

From the west, at Chelatchie, turn east from Highway 504 onto Healy Road (2-lane paved), which ultimately becomes FR 54. Continue up the Canyon Creek drainage for 8.9 miles to the intersection with FR 57. Turn uphill on FR 57 (1-lane paved), and in 3.5 miles meet the west end of FR 5704 (1-lane gravel). Follow FR 5704 for 2 miles across the south side of West Point to its intersection with FR 5704315.

Fire lookouts were placed to maximize the amount of countryside that could be seen, and this old lookout site was no exception. The cupola cabin that was built here around 1925 no longer exists, but the vast sweeping views down Canyon and Siouxon creek drainages remain. To the north, south, and east the tops of the volcanic giants of the South Cascades stretch the horizon: St. Helens, Adams, Hood, and the distant Jefferson.

Unfortunately, the scenery in the immediate vicinity is less appealing. The upper slopes and south side of the point have been stripped by clearcuts; the nakedness emphasizes the peak's distinctive, sharply pointed

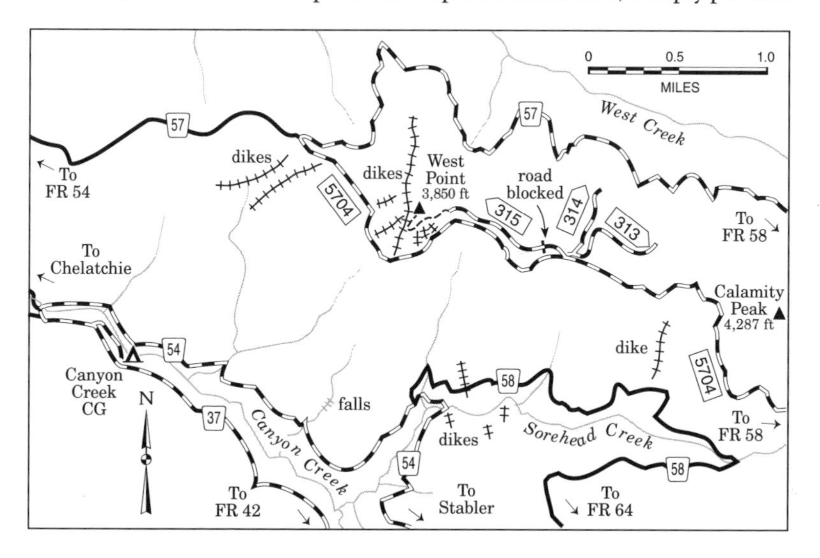

West Point was once the site of a fire lookout.

cone shape. Several long, narrow, linear rock outcrops, interesting features of the area's geology, are also exposed. These are hornblende-andesite dikes, formed between 20 and 15 million years ago when magma forced its way up through crack systems in the overlying rock structure.

Multiple dikes (technically, dike swarms) arrayed around and pointing at a single geographic feature generally indicate that the feature was either a volcano vent or an intrusive dome of lava originating from the same magma source as the dikes. Because the dikes here focus on the summit of West Point, it's likely that it was created by such volcanic activity.

To reach the top of West Point, either continue driving a short distance west on rough, narrow FR 5704315 to where the road is blocked by a jumble of rotting logs, or walk the road from its start point. Beyond the barrier the grade drops gradually, just below the ridge top, to a saddle east of the summit. From here it begins a gradual ascent across the south face of the point, deteriorates to a cat track on the south shoulder, and, after a switchback, meets the old trail to the lookout. Although the trail has long been abandoned, it is easily followed to the final few stairs cut in the rock that reach the lookout site, a few feet below the actual summit.

For yet another close-up look at a volcanic dike and the lava flows that created the local topography, continue east on FR 54 beyond the FR 57 junction for 4.6 miles to the junction with FR 58, then on beyond for 0.3 mile, where a roadcut shows a diverse combination of colors and shapes in the lava flows—green to orange to brown, platy to blocky to columnar. In a few hundred feet is Sorehead Creek and a close-up view of the rib of a volcanic dike protruding from the ridge just east of the creek. Similar dikes are hidden in the trees upstream.

5

The Indian Heaven Rift

A major zone of volcanic activity runs north–south through the heart of the Indian Heaven Wilderness. More than forty-five individual volcano vents have been identified in this area, ranging from older shield volcanoes to younger vents and cinder cones. The shield volcanoes periodically erupted large flows of olivine basalt between 800,000 and 450,000 years ago; the younger vents, which extruded basalt lavas, are less than 150,000 years old.

A counterclockwise driving loop that circles the perimeter of the Indian Heaven Wilderness provides access to many of its interesting geological features. The route described here enters the area from the southwest via Wind River Highway and in its circuit passes between the Indian Heaven Wilderness and the north side of Big Lava Bed.

To reach the starting point, drive Wind River Highway (2-lane paved) north from Highway 14 at Carson for 5.6 miles, to where a road signed to Panther Creek heads east. In 50 feet Panther Creek Road (1-lane paved, with pullouts) leaves to the north, changing from a county road to FR 65 in about 1 mile. The road passes the entrance to Panther Creek Campground in 2.8 miles, then continues steadily uphill through heavy forest and a succession of older clearcuts for another 8.5 miles to Four Corners, a junction with FR 60. All travel beyond this point, with the exception of a couple of miles on FR 24, is on 1½-lane gravel roads.

All hikes and cross-country travel described in this section can be made as one-day round trips, although some are fairly long. The Pacific Crest National Scenic Trail

A formation of pahoehoe lava in Big Lava Bed

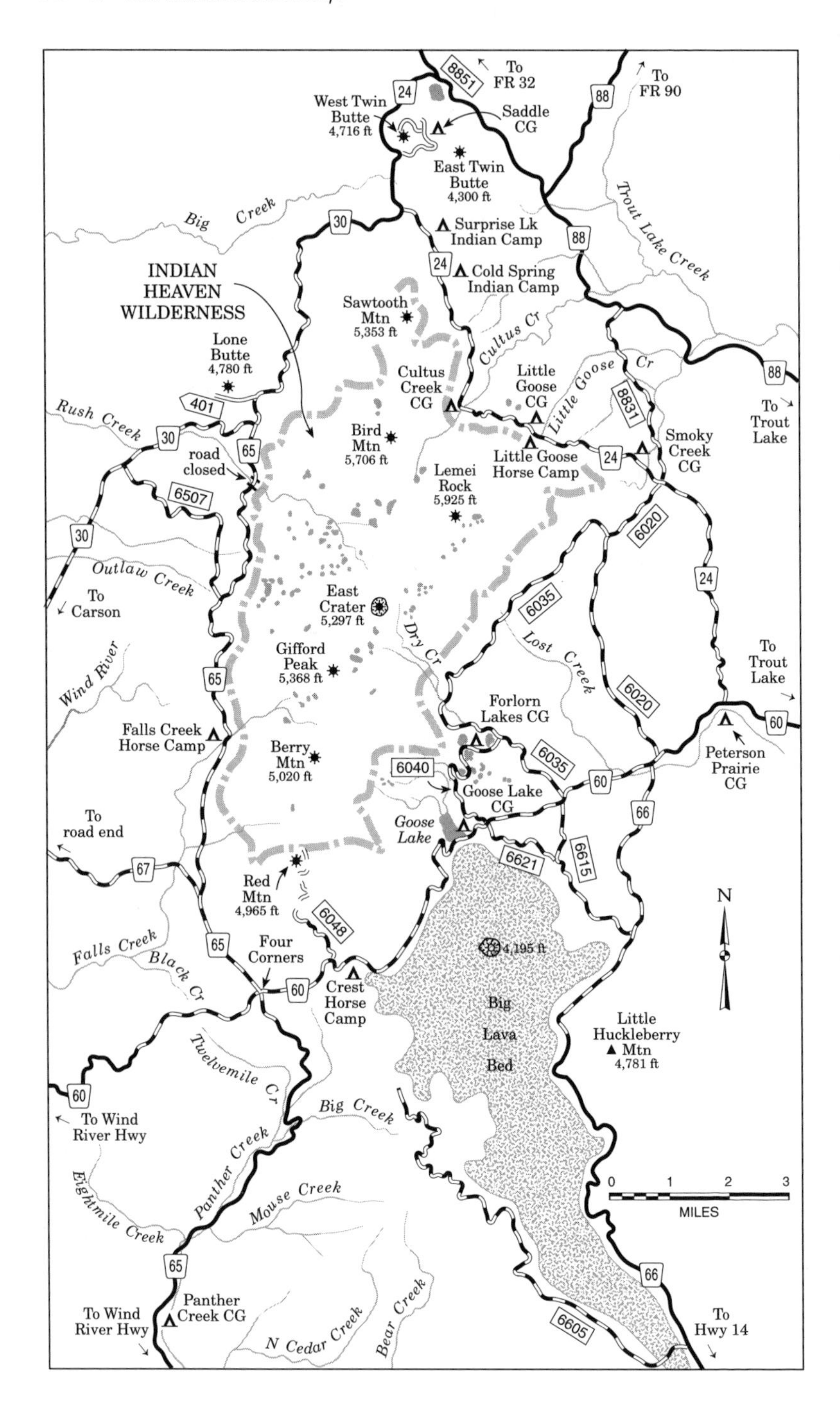
To FR 32
8851
To FR 90
88
24
West Twin Butte 4,716 ft
Saddle CG
East Twin Butte 4,300 ft
Big Creek
30
Surprise Lk Indian Camp
88
Trout Lake Creek
24
Cold Spring Indian Camp
INDIAN HEAVEN WILDERNESS
Sawtooth Mtn 5,353 ft
Cultus Cr
Lone Butte 4,780 ft
Cultus Creek CG
Little Goose CG
Little Goose Cr
88
401
8831
To Trout Lake
Rush Creek
Bird Mtn 5,706 ft
30
road closed
65
Little Goose Horse Camp
24
Smoky Creek CG
Lemei Rock 5,925 ft
6507
6020
30
24
Outlaw Creek
To Carson
East Crater 5,297 ft
6035
Dry Cr
Lost Creek
Wind River
Gifford Peak 5,368 ft
To Trout Lake
65
6020
Forlorn Lakes CG
Falls Creek Horse Camp
Berry Mtn 5,020 ft
6035
60
Peterson Prairie CG
6040
60
Goose Lake CG
To road end
Goose Lake
66
6621
6615
67
Red Mtn 4,965 ft
N
6048
4,195 ft
65
Falls Creek
Black Cr
Four Corners
60
Crest Horse Camp
Big Lava Bed
Little Huckleberry Mtn 4,781 ft
Twelvemile Cr
60
To Wind River Hwy
Big Creek
Panther Creek
0 1 2 3
MILES
Eightmile Creek
Mouse Creek
65
66
Panther Creek CG
To Wind River Hwy
Bear Creek
N Cedar Creek
6605
To Hwy 14

runs north–south through the wilderness. Because the PCT is typically done with one or more backcountry overnight camps, and requires prearranged transportation at either end, it is not described here in its entirety; however, some of the trips described here make use of sections of the PCT.

Red Mountain

Road trip to the top of a cinder cone

Elevation: Junction of FR 60 and FR 6048 3,370 feet, summit 4,965 feet
Maps: USGS Gifford Peak, USFS Mount Adams Ranger District
Driving Directions: At the Four Corners junction of FR 65 and FR 60, turn east on FR 60 (1½-lane gravel), and in 1.6 miles head north on FR 6048 (1-lane dirt). Vehicles with adequate clearance, and preferably four-wheel-drive, can make it to the summit tower when the gate is open. Otherwise, hike from wherever you decide to park along the 4-mile road.

As you leave Four Corners on FR 60, the most dramatic point on the skyline is Red Mountain, with clearcuts and sidehill forests sweeping up to its steep, barren, 600-foot-high summit cone. (No doubt about its identity—it's covered with red cinders.) The fire lookout on its summit is one of three remaining in the Gifford Pinchot National Forest that are still actively staffed. Views from the lookout site to every point of the compass are spectacular!

The road to the lookout is gated, except in summer when the tower is staffed. Even when open, this is definitely not a route for the average passenger car; a sign at the start proclaims "End of Maintenance." In a mile another warns of a "Primitive Road" beyond. This is no understatement, because the road rapidly becomes narrow, with ruts, chuckholes, dust when dry, and slick mud when wet. A slim, exposed section of road above the third major switchback crosses a loose, pea-gravel cinder bed, with extremely poor traction. You may be wise to park and hike the road at some point before your nerves and your vehicle's capabilities are pushed too far.

The last switchback below the summit marks the southern boundary of the Indian Heaven Wilderness and the start of Trail #171 to Indian Race Track. The final leg to the top of the mountain passes an old, weathered shed and a helipad just below the tower. The lookout welcomes visitors between 9:00 A.M. and 5:00 P.M.; respect privacy at other times.

Unending views stretch for over 50 miles in any direction. To the north the spine of the Cascade Range traces through forest-covered slopes that have overgrown the shield volcanoes of Berry Mountain and Gifford Peak. Behind them is the flat top of East Crater, an unimposing summit that conceals a deep crater core. The snowy cone of Rainier floats in the distance. To the northeast is the cliffy outcrop of Lemei Rock, below which is Lake Wapiki, the progenitor of the Ice Cave basalt lava flows that extend east for nearly 10 miles to Trout Lake. The ice-clad mass of Mount Adams

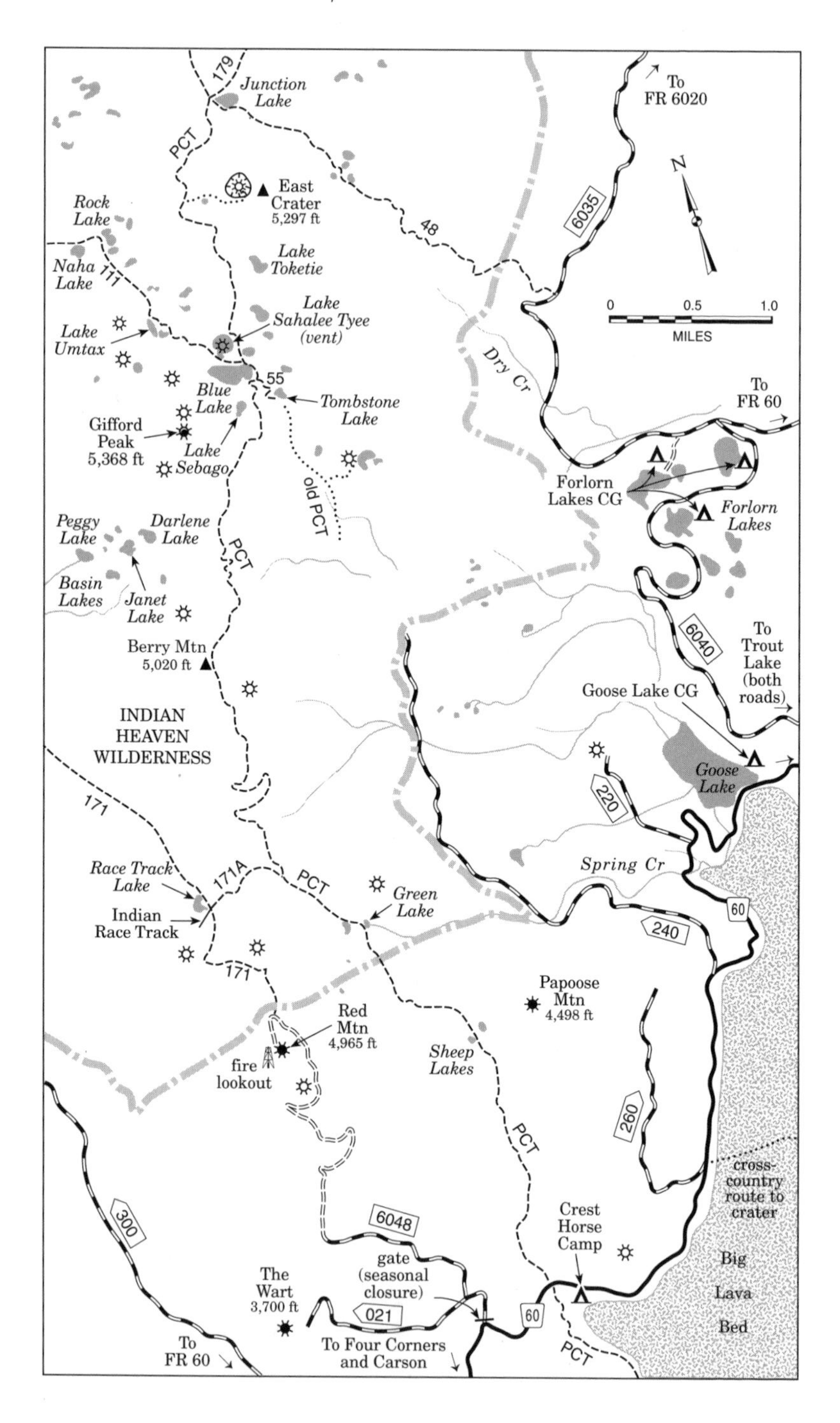

Junction Lake
179
PCT
To FR 6020
East Crater 5,297 ft
Rock Lake
48
6035
N
Naha Lake
111
Lake Toketie
Lake Sahalee Tyee (vent)
0 0.5 1.0
MILES
Lake Umtax
Dry Cr
55
Blue Lake
Tombstone Lake
To FR 60
Gifford Peak 5,368 ft
Lake Sebago
old PCT
Forlorn Lakes CG
Forlorn Lakes
Peggy Lake
Darlene Lake
PCT
Basin Lakes
Janet Lake
6040
To Trout Lake (both roads)
Berry Mtn 5,020 ft
Goose Lake CG
INDIAN HEAVEN WILDERNESS
Goose Lake
220
171
Spring Cr
Race Track Lake
171A
PCT
Green Lake
Indian Race Track
60
240
171
Papoose Mtn 4,498 ft
Red Mtn 4,965 ft
fire lookout
Sheep Lakes
260
PCT
cross-country route to crater
300
6048
Crest Horse Camp
Big Lava Bed
gate (seasonal closure)
The Wart 3,700 ft
021
60
To FR 60
To Four Corners and Carson
PCT

Red Mountain lookout at sunrise. The stratovolcano of Mount Hood, in Oregon, is in the distance.

rises above the ridgeline that marks the limits of the flow. To the northwest another prominent volcano shape, Lone Butte, rides the crest of the intervening sea of timber. The blasted summit of St. Helens lies low on the horizon.

Downslope to the southwest is The Wart, a low, wooded neighbor—yet another volcano vent that is newer than the initial eruptions from the one on which you stand. In the foreground to the south, the forest suddenly changes color from dark green to lighter, drab gray-green of stunted pine that marks the location of Big Lava Bed. The darker mass with a scooped-out top near the north end of the lava bed is the crater at the source of the flow. Three cone-shaped summits along its western edge, partially shorn of trees by logging, are volcanoes of a more ancient age.

Red Mountain began as a shield volcano, erupting several layers of olivine basalt that accumulated to a thickness of 50 to 100 feet over a period between 240,000 and 170,000 years ago. The earliest lavas flowed mainly south and southwest into the area between Goose Lake and Falls, Black, and Twelvemile creeks. Later cinder cones, probably some 150,000 years old, form the presently recognizable contours of Red Mountain and its lower, unnamed companion cone immediately to the north. These cones ejected blocky basalt that spread southeast. The flow banked against an earlier vent at Papoose Mountain, which overlays earlier Red Mountain flows in the area between Goose Lake and the present route of the PCT.

Berry Mountain and Gifford Peak

Hiking trails to volcano vents

Trails: PCT #2000, Race Track Trail #171, Shortcut Trail #171A
Rating: (M); hikers, saddle and pack stock
Distance: Via #2000 6.1 miles, via #171 and #171A and #2000 4.5 miles
Elevation: Via #2000: FR 60 trailhead 3,490 feet, top of Berry Mountain 5,020 feet, Basin Lakes saddle 4,730 feet; via #171: FR 60048 trailhead 4,800 feet
Maps: USGS Gifford Peak, USFS Mount Adams Ranger District
Driving Directions: Take either the Red Mountain Lookout road, FR 6048 (1-lane dirt), preceding, to the start of Trail #171 to Indian Race Track, or follow FR 60 (1½-lane gravel) east from Four Corners for 2 miles to Crest Horse Camp, where #2000 crosses the road.

Hike across the top of several dormant volcano vents with dazzling views of the South Cascades stratovolcanoes. Closer at hand are the ghosts of other volcanoes within the Indian Heaven Rift: Red Mountain, Lemei Rock, East Crater, and Gifford Peak. This trek arbitrarily ends at the saddle between Berry Mountain and Gifford Peak, with tantalizing glimpses down to the lakes and tarns, collectively known as the Basin Lakes, that lie in the glacier-carved cirque to the west. The outing could easily be extended to a comfortable overnight backpack by continuing north for another mile to the topaz-colored pool of Blue Lake, tucked into the northeast slope of Gifford Peak—perhaps the most beautiful campsite in the Indian Heaven.

The shortest of the two options for this trip, from Red Mountain, has ups and downs both coming and going. The longer of the two options, hiking the whole way via the PCT, is all uphill going in and all downhill going out. Let's try the Red Mountain version first. If your vehicle is capable of handling the nasty conditions on the upper extent of FR 6048, park at the last switchback below the summit, fill out a wilderness hiking permit, and hike gradually downhill through hemlock and fir to Indian Race Track, an open meadow at the edge of a shallow pond.

The meadow blazes with wildflowers in late summer and fall. The Race Track is little more than a deep groove, now being reclaimed by ground cover, that runs northeast–southwest through the meadow before disappearing into an encroaching growth of young fir. In the late 1800s and early 1900s, this was a large meadow where the Klickitat and Yakama tribes gathered regularly to harvest huckleberries, socialize, and engage in horse races. These frequent races wore a permanent indentation in the turf. Much of the former meadow is now overgrown by forest.

From Indian Race Track a short, almost level path leads a mile east to join the PCT. North from this junction the trail soon climbs steeply through three long switchbacks to near the top of Berry Mountain. Between 300,000 and 130,000 years ago the mountain was created from a series of volcano vents along its crest. North of Berry Mountain is Gifford Peak, the oldest volcano in the Indian Heaven Rift. Three to five vents along the summit first extruded basalt lavas over 730,000 years ago. They laid down a shield of lava some 8 miles in diameter and 5 to 20 feet deep around the eruptive vents. The final stages of volcanic activity built the cinder cones that now make up Berry Mountain and Gifford Peak. There has been no volcanic activity from these sources for over 100,000 years.

From the saddle between these two mountains, look down upon the Basin Lakes to the west. These nine lakes and several more tarns lie in a deep cirque that was carved by local alpine glaciers between 25,000 and 17,000 years ago. Hikers once could descend west through the basin to intersect Trail #171 near FR 65 via a way-trail; today, however, only an experienced cross-country navigator can find segments of the path.

For the second route option, following the PCT for the entire length of the trip, start from Crest Horse Camp where it crosses FR 60. A plush campsite? No—it's only a pit toilet, a few picnic tables and fire grates, and no water. From the camp the trail climbs gradually but relentlessly north along a sometimes-flowing creek that drains two tiny tarns called Sheep Lakes. The forested rise east of the lakes covers yet another volcano vent that dumped lava southward for a few miles some 130,000 years ago. The PCT continues its gradual ascent to Green Lake, a pond on the western fringe of yet another young lava flow. The route continues northwest for another 0.5 mile to the junction with Trail #171A.

Big Lava Bed

Cross-country route to a volcano crater

Rating: (D); hikers
Distance: From FR 60 to Big Lava Bed crater 2 miles
Elevation: Starting point on FR 60 3,540 feet, crater rim 4,195 feet
Maps: USGS Big Huckleberry Mountain, Gifford Peak, Little Huckleberry Mountain, Willard; USFS Mount Adams Ranger District
Driving Directions: From Four Corners head west on FR 60 (1½-lane gravel) for 2.4 miles, where the road parallels the northwest side of the lava bed for another 4 miles to Goose Lake. Another road, FR 66, skirts the east side of the lava bed. To reach it, 10 miles east of Carson turn north from Highway 14 on the road signed to Willard and the Little White Salmon Fish Hatchery (2-lane paved). Continue north for 7.5 miles past Willard, then turn northwest on FR 66 (2-lane paved that becomes gravel). In 2 miles this road passes the south end of Big Lava Bed then continues along the east side of the bed for 11.2 miles to FR 6615 (1-lane gravel). Take FR 6615 northwest for 2 miles to FR 6621 (1-lane gravel), then follow it northwest for 1.3 miles to meet FR 60, 0.4 miles east of Goose Lake.

This is one of the most primitive, untracked, least-explored, and least-known areas remaining in the South Cascades. The lava bed covers an area of over 20 square miles, running south from FR 60 for 10 miles to Willard, 5 miles north of the Columbia River. This extremely young flow erupted only 8,200 years ago. When seen from surrounding mountains, it is distinguishable only by the contrast of lighter gray-green of its sparse pine forest cover, which contrasts with the deeper green of hemlock and fir in the surrounding forests. Up close the lava flow presents a different and more complex picture. Viewed from the PCT and forest roads that skirt its perimeter, the flow is a massive jumble of pahoehoe basalt flows fractured into a maze of huge lava blocks, fissures, pressure ridges, and crevices, all topped with a uniform growth of small pine—a difficult area to traverse and a place where one can easily get lost!

The source vent for this flow is found near its north end, about 1.5 miles south of Goose Lake. Here a complex, 200-foot-high platform of lava, laced with 50-foot-high hummocks, narrow channels, and collapsed lava tubes, rises east to the 800-foot-high rim surrounding the crater that fed the flow. Most of the lava burst from the northwest side of the crater rim then swept south around its flanks. The eruption consisted of one or more thin (5 to 30 feet thick) layers of fluid basalt that formed narrow lava tubes as it spread south. Instead of creating lava tube caves, as was the case with thicker flows to the north and on St. Helens, when the flow surfaces here began to congeal, pressure from still-molten lava beneath lifted the newly formed lava skin, forming a platform west of the vent. This subsurface inflation fractured the surface into a complex array of ridges, small mesas,

This aerial photo shows the crater at the heart of Big Lava Bed.

and deep fissures. At the southern end of the flow the volcanic gasses also ballooned the lava skin to form hollow bulbs (tumuli) just below the surface.

Most visitors are content to examine the surrealistic lava formations that can be seen within a hundred feet of the road. No trails cross its expanse, although there are reports of two primitive paths that supposedly crossed the lava flows before the days of the Forest Service.

Exploring the crater is fascinating but very challenging. The shortest approach is from the west, starting from the junction of FR 60 and FR 6000260. The crater is about 1.5 miles due east of this junction and is frequently visible through the cover of stunted pine that covers the lava bed. The rough scramble east rarely follows a straight line because detours are forced by hodgepodge piles of basalt blocks and deep, narrow fractures in the lava surface. One short, steep, 150-foot-high face is surmounted before the floor levels to a more gradual ascent to the west crater rim. Once atop the heavily wooded rim, the slope tapers gently down about 100 feet into its heart then drops into a tiny meadow that becomes a miniature flower garden after winter snows disappear.

Near the south end of the lava flow, cross-country exploration can reveal spots where the lower ends of basalt tubes expanded rapidly forming hollow bulbs that fractured, opening like egg shells. Over the years the interiors of these bulbs have become host to a complex, varied, and sometimes rare collection of biota.

Goose Lake

Road trip to a lake formed by lava flows

Elevation: 3,100 feet
Maps: USGS Gifford Peak, USFS Mount Adams Ranger District
Driving Directions: From Four Corners, head east on FR 60 (1½-lane gravel) for 6.4 miles to Goose Lake Campground.

Ghostly silvered spines of dead trees rise from the waters of Goose Lake and stud its shoreline, projecting an aura of mystery to the surroundings. How did this dead forest come to be? Goose Lake was created about 8,200 years ago when flows from the Big Lava Bed volcano blocked stream drainages from the northwest. Once formed, the water level varied periodically as it filled with spring run-off, then drained through an outlet hole in the middle of the lake. The outlet was dammed by man in the 1930s, raising the level of the lake by several feet and drowning the shoreline forest. The silvered stumps remain.

The sense of the supernatural is reinforced by an Indian legend and pioneer tales about a set of hand- and footprints that, in the 1890s, were found impressed in shoreside lava. According to the natives these prints, called Wa-ti-kch, belonged to a young woman who leaped from either Bird Mountain or Lemei Rock to escape the pursuit of an evil spirit. The legend

Goose Lake was created over 8,000 years ago when stream drainages were blocked by lava flows.

maintains that in the evening the woman can sometimes be seen sitting on a rock beside the lake, combing her hair.

Some who saw the prints swore they were actual imprints that had been pressed into molten lava; others were just as sure that they had merely been carved into the rock. In an article written in 1919, a well-known geologist who examined them asserted they were actual prints. He described them as being about 2 feet apart and impressed between ¼ and ⅞ inch into the lava rock. The footprints were moccasined, and the fingers of the handprints were widely spread. Although the prints were inundated when the lake was dammed in the 1930s, the lake periodically drained, for no discernible reason, between then and the 1960s, and the prints would again be seen. Suddenly, the outlet plugged up permanently, and in ensuing years record of the location of the prints was lost. In recent years a Portland man snorkeled in the lake in search of the prints. Coming to a likely looking spot, he brushed aside algae on the bottom and found them!

A campground with limited facilities is located on the hillside along the east side of the lake; a gravel boat launch ramp enters the lake near the campground entrance. Weird, massive, contorted piles of pahoehoe lava, part of the flows that initially formed the lake, can be found across the road to the south.

East Crater

Hiking trails and cross-country route to a volcano crater

Trails: Trail #48, PCT #2000
Rating: Trails (M), cross-country (D); hikers, saddle and pack stock
Distance: FR 6035 to cross-country start point 3.1 miles, cross-country 0.3 mile
Elevation: FR 6035 trailhead 4,080 feet, crater rim 5,100 to 5,297 feet, bottom of crater 4,070 feet
Maps: USGS Gifford Peak, Lone Butte; USFS Mount Adams Ranger District
Driving Directions: Continue east from Goose Lake on FR 60 (1½-lane gravel) for 2 miles, then turn northeast on FR 6035 (1-lane gravel). Follow the road past Forlorn Lakes Campground, and in 4 miles reach the start of Trail #48.

What may appear to some as yet another innocuous wooded hill harbors an internal secret—the deep crater bowl of an old cinder cone. No trail reaches this hidden spot; a stiff, 0.5-mile cross-country trek is required to reach the secluded center of this old volcano vent.

The shortest approach is via Trail #48, which leaves FR 6035, then climbs gradually northeast through mountain hemlock to a meadow with a scattering of tarns. Here a profusion of early summer wildflowers and the tasty huckleberries of fall tempt hikers to dally rather than pursue more esoteric goals. Beyond these meadows the path enters alpine growth of noble and true fir, then reaches the shore of Junction Lake, the intersection of Trails #48, #179, and #2000 (the PCT).

Follow the PCT south as it skirts the north and west sides of East Crater. At the trail's high point, due west of the East Crater rim, head cross-country on a gradual ascent to the southeast. A spring-fed pond on the east shoulder of the crater marks the site of a vent that erupted flows of platy olivine basalt that spread southwest across the older Red Mountain flows some 30,000 to 25,000 years ago. To avoid scattered rock cliffs that frame the northernmost two-thirds of the rim, continue an uphill traverse to the southeast, then east, to the low point of the crater rim.

From this saddle, follow animal trails downhill north to the open meadow, the size of a football field. Scattered blocks of orange-red cinders litter the beargrass-covered crater floor. On all sides steep, wooded slopes encircle what was once a caldron at the heart of an erupting volcano.

Spooky indeed, because on a geological time scale the eruptions from this vent may not be over—just temporarily suspended. Carefully retrace your steps and, with the aid of a compass course to the west, rejoin the PCT for the return trip.

Lake Sahalee Tyee

Hiking trails and cross-country hiking route to volcano vents

Trails: East Crater Trail #48, PCT #2000, Thomas Lake Trail #111
Rating: (M); hikers, saddle and pack stock
Distance: Via #48 and #2000 4.5 miles, via #111 3.3 miles
Elevation: Trailhead of #48 at FR 6035 4,080 feet, trailhead of #111 at FR 65 4,040 feet, Lake Sahalee Tyee 4,700 feet
Maps: USGS Gifford Peak, USGS Mount Adams Ranger District
Driving Directions: From the east, see East Crater, preceding. From the west, take FR 65 (1½-lane gravel) 8.7 miles north from Four Corners to the start of Trail #111.

This small, deep, almost perfectly round lake rarely draws the crowds or the raves that its nearby companion, Blue Lake, does. The latter is larger, lies at the base of picturesque cliffs on the north side of Gifford Peak, and is a gem-like, deep azure. But what Sahalee Tyee does have, in addition to its lyrical Indian name, meaning Great Spirit, is its unique geological origin. The shallow bowl in which the lake lies is the crater of an old volcano. Not a large one by Indian Heaven standards, but a reasonably young one, less than 130,000 years old. The volcano erupted a platy olivine basalt that spread southeast into the headwaters of Dry Creek.

To reach Sahalee Tyee from the east, follow the directions to East Crater in the preceding hike, but instead of leaving the PCT at the saddle west of the crater, continue south on the trail another mile, wandering gradually down through forest to the junction with Trail #111 and meadows surrounding Blue Lake. Turn west and follow the shore for 0.2 mile to Sahalee Tyee.

A shorter route to the lake, and also the most picturesque, is from the west via Trail #111; this is also the more physically demanding route. From FR 65 an easy grade slips through huckleberries and, with a few gentle switchbacks, reaches a narrow thread of meadow and forest between Dee and Heather lakes to the north and larger Thomas Lake to the south. Boot paths lead along the shore of Thomas Lake to Lake Kwaddis, just a step or two beyond its south end. Anglers' paths lead east from Heather to the pocket holding Eunice Lake.

A short, steep ascent gains a flat parkland. A few hundred feet cross-country to the south is Lake Le-Loo. The path continues east through meadows, then angles south and climbs to skirt the top of a talus slope before reaching the west shore of tiny, shallow Naha Lake. A few hundred yards farther the route turns abruptly south, but straight ahead is Rock Lake; it gains its name from boulders that break the surface along its meadow-edged shore. The trail now wanders through a plateau meadow, fringed with low fir, and dappled by a myriad of small ponds and lakelets, then finally arrives at Lake Sahalee Tyee and its beautiful companion, Blue Lake.

A mile downhill, to the southeast of Sahalee Tyee, are another pair of unnamed lakes that also fill a one-time crater, this one even younger than Sahalee Tyee. When this vent broke through the Sahalee Tyee flow, about 5,000 years after the earlier eruption, the lava spread on top of the previous flow to reach and cover the Forlorn Lakes basin. To get to this crater and its lakes, take Trail #55 from Blue Lake to Tombstone Lake, and locate the old route of the PCT south from Tombstone Lake. Follow it for 0.5 mile, then head cross-country to the east for 0.2 mile to the lake-filled crater.

Lemei Rock and Lake Wapiki

Hiking trails to volcano vents

Trails: Indian Heaven Trail #33, Lemei Trail #34, Lake Wapiki Trail #34A
Rating: #33 (M), #34 (D), #34A (E); hikers, saddle and pack stock
Distance: Lemei saddle: via #33 4.5 miles, via #34 4.1 miles; Lake Wapiki: via #33 5 miles, via #34 3.6 miles
Elevation: #33 trailhead 3,988 feet, #34 trailhead 3,650 feet, Lemei saddle 5,600 feet, Lake Wapiki 5,230 feet
Maps: USGS Lone Butte, Sleeping Beauty; USFS Mount Adams Ranger District
Driving Directions: Continue east from Goose Lake on FR 60 (1½-lane gravel) for 3.8 miles, then turn north on FR 6020 (1-lane gravel) and follow it for 5.5 miles to FR 24 (1-lane gravel). Head northwest on FR 24 to either the start of Trail #34, 0.2 mile, or to Cultus Creek Campground, 4.3 miles.

One of the most dramatic features in the Indian Heaven is the craggy, 300-foot-high slab of Lemei Rock, the highest point in the wilderness. The summit stares down into the bowl of a lower crater wrapped around the cobalt-blue waters of Lake Wapiki. As Trail #34 crosses the bare, northwest shoulder of the rock, its blocky summit towers seem only a touch away. South along the Indian Heaven crest is the snaggled top of Sawtooth Mountain. Northwest, the sharp, rocky tooth of Sleeping Beauty thrusts into the foreground below the ice-capped mass of Mount Adams, and the rugged backbone of the Goat Rocks lines the distant horizon.

Lemei Rock, a vent on the ridge to the southeast, and the crater now filled by Lake Wapiki are the sources of some of the most extensive lava flows to issue from the Indian Heaven Rift. Eruptions that occurred more than 300,000 years ago from a volcano at Lemei Rock overlaid earlier Bird Mountain lava fields. The core of the vent that discharged this lava, which was extensively eroded by subsequent glaciation, comprises Lemei Rock.

Lemei Rock is the core of an old volcano that erupted 300,000 years ago.

Around 30,000 years ago a vent on the ridge to the southeast spread new basalt around the base of the old core and downslope to the southeast. Sometime within the next 6,000 years, another magma column broke through overlying volcanic rock about ½ mile east of the old core at the present site of Lake Wapiki, and a series of thin, very fluid tongues of basalt lapped through a breach in the crater wall, swept east down the valley as far as Trout Lake, then angled south and continued down the White Salmon valley to beyond Husum. Cumulatively, these flows reached depths of up to 100 feet and spread as far as 45 miles from the source crater.

Several routes through the wilderness can be used to reach Lemei Rock; the two shortest are from either the north via Trails #33 and #34 or from the east via Trail #34. Your choice depends on your objective. If the scenic view from the high point on the northwest rib of Lemei Rock is your goal, either route is about 4¼ miles. However, the hike from the north gains 1,610 feet of elevation, while the one from the east gains 1,950 feet. The north is the winner.

If headed for Lake Wapiki, the route from the north totals 5.8 miles, with an elevation gain of 1,840 feet, a loss of 600 feet (with the high-point

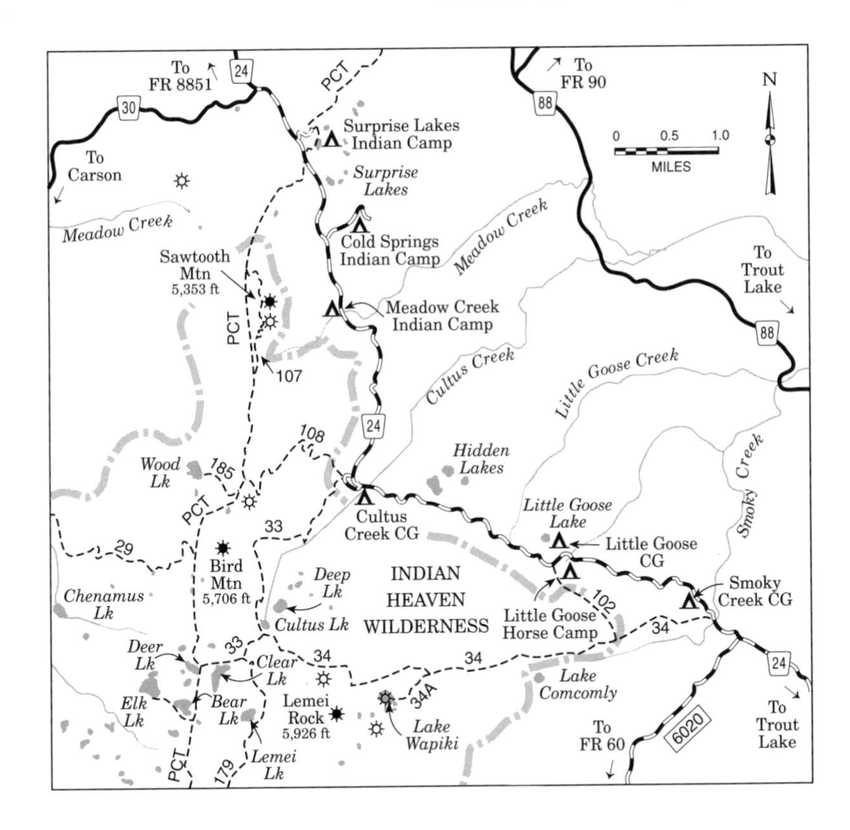

scenery tossed in as a bonus), and a gain of 230 feet to the lake; the east route is 3.6 miles with a gain of 1,580 feet. Decision: east approach. The best of all possible worlds is a combination of hikes starting from the north on Trail #33 and coming out to the east on Trail #34. Prearrange transportation at both ends. This 9.6-mile trip includes a visit to Lake Wapiki.

The northern route, via Trail #33, leaves Cultus Creek Campground on an uphill grind that becomes even more difficult in about ¾ mile as it struggles up the lip of a wooded cirque near Cultus Creek. Once atop this obstacle, the old path has since been rerouted to the west to make the grade easier, albeit slightly longer. The route turns south as it skirts below the base of Bird Mountain, occasionally breaking out of forest cover for glimpses of cliff faces above. At around 5,100 feet reach small meadows that burst with wildflower color in early summer. The route passes Trail #33A, a short spur that descends east to Deep Lake, then wanders along the west shore of Cultus Lake to the junction with Trail #34. The latter route climbs gradually east through flowered parkland to a magnificent, rock-bound saddle viewpoint below Lemei Rock. The ascent to the summit requires rock-climbing skills. Continue the hike to Lake Wapiki by dropping to Trail #34A and taking it uphill to the lake.

The eastern route starts its ascent on Trail #34 from FR 24, southeast of Smoky Creek Campground. The higher it gets, the steeper it becomes, as the path climbs over old volcanic flows, now covered with timber. Early season wildflowers give way to trailside huckleberries in fall; the common ingredient through all seasons (except late fall) is a cloud of mosquitoes. Tiny meadow breaks reveal views north across the face of Sleeping Beauty to glaciers flowing from the summit of Mount Adams. In just over 3 miles Lake Wapiki Trail #34A bears southwest. In a scant 0.5 mile arrive at the east shore of this blue, blue lake, enveloped on all sides by steep, black, blocky basalt slopes, half hidden in trees, that mark the rim of this one-time volcanic cauldron.

Bird Mountain

Hiking trail past volcano vents

Trails: Indian Heaven Trail #33, PCT #2000, Cultus Creek Trail #108
Rating: #33 and #2000 (M), #108 (D); hikers, saddle and pack stock
Distance: 6.8 miles
Elevation: Cultus Creek Campground 3,988 feet, high point at junction of #2000 and #108 5,237 feet
Maps: USGS Lone Butte, USFS Mount Adams Ranger District
Driving Directions: Continue east from Goose Lake on FR 60 (1½-lane gravel) for 3.8 miles, then turn north on FR 6020 (1-lane gravel) and follow it north for 5.5 miles to FR 24 (1-lane gravel). Head northwest on FR 24 to Cultus Creek Campground, 4.3 miles

The highlight of this loop trail is eye-popping views of every major volcanic summit within a 50-mile radius, as well as a host of smaller intermediate cinder cones and lava intrusions. The hike around one of the older groups of lava vents in the Indian Heaven Wilderness is challenging; Trail #108 gains 1,250 feet in only 1.5 miles, but the outstanding scenery more than compensates for the effort.

Bird Mountain is a contorted, 1.5-mile-long backbone ridge of the Cascade Crest through the Indian Heaven Wilderness. Either two or four volcano vents form the ridge, depending on which geologist's opinion you choose. Most agree, however, that massive, widespread, pahoehoe flows of olivine basalt issued from the dikes on the west side of Bird Mountain somewhere around 130,000 years ago and spread over the northwest portion of the wilderness, covering older accumulations of lava and volcanic debris to a depth of 5 to 30 feet.

Your choice of direction for the loop trip depends on how well your knees cope with uphill versus downhill strain. A clockwise route gains altitude more slowly but comes downhill with a joint-jarring steepness. Counterclockwise, the trail is a stiff uphill, but the return route is longer and more gentle.

This description follows the clockwise path. The section of Trail #33 between Cultus Creek Campground and the junction with Trail #34 at Cultus Lake is described in the Lemei Rock and Lake Wapiki hike, preceding. Beyond the Trail #34 junction the path winds downhill west, passes the north end of Trail #179, and enters meadows below a rockslide at the north end of Clear Lake. Another 0.2 mile west the trail ends at its intersection with the PCT on the east shore of Deer Lake.

North from Deer Lake, the PCT climbs to a sidehill traverse, passes a couple of ponds, then slips along the west side of Bird Mountain. After skirting the shore of another meadow-rimmed lakelet, the path ascends past Trail #185 to Wood Lake, then meets Trail #108 just below a saddle in the long Bird Mountain summit ridge. Head east on this trail.

As Trail #108 reaches the saddle above the bare rock face on the east side of Bird Mountain, awesome views are exposed of glacier tongues lapping down the massive west slope of Mount Adams. Foreground ridges are topped by the rock promontories of Steamboat Mountain and Sleeping Beauty, and the Goat Rocks line the northwest horizon. North, the vertical rock face of Sawtooth Mountain looms close at hand, while Rainier crowns the skyline.

A short side trip south along the ridge crest leads to Point 5,568, one of the old volcano vents of Bird Mountain. Here, Hood and St. Helens add to the list of scenic splendors. The true summit of Bird Mountain lies southwest at the end of a rough, ridge-crest scramble that doesn't justify the effort with improved views.

The path now drops east across the rocky, heather-clad face of the mountain, the brutally steep descent mitigated by only one long switchback. After a loss of 600 feet of elevation, the way enters meadows framed by clumps of small fir and hemlock. Once within the woods, more switchbacks snake down through rock outcrops, and after a brief respite the trail drops to Cultus Creek Campground.

Sawtooth Mountain

Hiking trail past volcano vents

Trails: PCT #2000, Sawtooth Trail #107
Rating: (M); hikers, saddle and pack stock
Distance: 5.8 miles
Elevation: FR 24 trailhead 4,238 feet, #107 high point 5,250 feet
Maps: USGS Lone Butte, USFS Mount Adams Ranger District
Driving Directions: Follow directions to Bird Mountain, preceding, as far as Cultus Lake Campground. Continue north on FR 24 (1-lane gravel) for 3.3 miles to the trailhead of #2000.

The ragged, rocky arête forming the top of Sawtooth Mountain is the most conspicuous feature in the north end of the Indian Heaven Wilderness. The east face drops away precipitously in a 500-foot-high vertical wall, but even from the west the snaggled upper 100 feet of the ridge stands out strikingly above surrounding forest. Sawtooth is one of the older Indian Heaven shield volcanoes. It flooded the area north and northwest with blocky olivine basalt around 300,000 to 240,000 years ago.

At one time the PCT ran along the top of the mountain just west of its crest; the trail was relocated in the 1970s to skirt its lower west flank, and the old section of trail higher up was redesignated Trail #107. A combination of the two trails makes an excellent one-day loop trip.

The PCT leaves FR 24 in the heart of the Sawtooth huckleberry fields. Berries east of the road are reserved for picking by Indians, while those to the west, through which the trail passes, are open to all comers. However, a Forest Service permit is required before harvesting berries. In about ½ mile the route climbs out of the berry fields into forest.

At a fork in 1.3 miles, take the east leg, Trail #107, which climbs gradually, but steadily, through several easy switchbacks to reach the base of the jagged mountain-top ridge. Here it levels and begins a 0.2-mile traverse along the west side of the mountain, about 100 feet below the ridge crest. Enjoy great views here of Rainier, St. Helens, and the nearby, distinctive volcano cone of Lone Butte. The adventuresome can scramble up between ridge-top teeth to the 5,323-foot-high summit of Sawtooth Mountain for even wider views that include the Goat Rocks, Sleeping Beauty, and Adams—and a nervous glance down the sheer southeast face.

After reaching the south end of the mountain, a few steeper switchbacks descend to the saddle between Sawtooth and Bird mountains, where the PCT is rejoined. Hike the gentle, wooded hillside north past the start of Trail #107 and on back to the FR 24 trailhead (perhaps hesitating on the way to feast on a handful or two of huckleberries).

Twin Buttes

Road trip and cross-country hike to the top of a volcano cone

Elevation: South Campground 4,030 feet, summit of West Twin Butte 4,716 feet

Maps: USGS Lone Butte, Sleeping Beauty; USFS Mount Adams Ranger District

Driving Directions: See Bird Mountain, earlier, for directions to Cultus Creek Campground. From there follow FR 24 (1-lane gravel, changing to 1-lane paved) north for 5.1 miles to FR 2480 (1-lane gravel), signed to South Camp and Saddle Camp. Head east, then north, on FR 2480 for 1 mile, then turn west on FR 2480031 (1-lane dirt), which circles the mountain to the summit of West Twin Butte.

Twin Buttes, two large, cone-shaped hills on the northern fringe of the Indian Heaven Wilderness, are relatively old additions to the local topography. Somewhere between 277,000 and 132,000 years ago a pair of magma vents burst through the older overlying strata of the Sawtooth Mountain shield volcano, erupting gas-bloated basalt cinders that built a pair of volcano cones to more than 700 feet above the local landscape. Today both are heavily timbered; only their soil and distinctive shapes attest to their birth.

Trails in the area are few; the PCT winds up the valley between the two cones, and traces of an old abandoned trail may be found crossing the south side of East Twin Butte. A forest road, primitive, rough, and challenging, climbs to the top of West Twin Butte where a fire lookout once stood; the structure burned in 1963.

Views from the old lookout site include the horizon-filling volcanic bulks of Rainier, Adams, and St. Helens. East are companion East Twin Butte and another old lookout site, Steamboat Mountain; to the south Sawtooth Mountain marks the end of the Indian Heaven Wilderness. Southwest is the solitary cone of Lone Butte.

Lone Butte

Road trip to a volcano cone

Elevation: Junction of FR 30 and FR 3000401 3,792 feet, Lone Butte quarry 4,120 feet, Lone Butte summit 4,780 feet

Maps: USGS Lone Butte, USFS Wind River Ranger District

Driving Directions: See driving directions to Twin Buttes, preceding. From its junction with FR 24 (1-lane paved with pullouts), drive south on FR 30 (1-lane paved, changing to 1½-lane gravel) for 5.9 miles to FR 300401 (1-lane gravel).

Lone Butte is an outstanding example of a uniquely formed volcano cone, but you will find no hiking trails here. Exposed strata in a gravel pit in the side slopes attest to its origins; a cross-country hike to the summit will provide both scenic views and mute evidence of its eruptive past. Lone Butte is clearly an old volcano cone when viewed from any quarter; a basalt core rises through trees at the summit, and impressive lava dikes, conspicuous from the north from FR 30, have been injected upward through its northeast slopes.

Lone Butte and other volcano vents to the northwest in the Crazy Hills are examples of *tuyas*—volcanoes that initially erupted beneath either glacial ice or meltwater lakes, built a cone through them, and then added to their height with additional lava flows above the glacier or lake surface. About 315,000 to 300,000 years ago the magma that created Lone Butte broke through glaciers or lakes that filled the Lewis River and the north end of the Indian Heaven. The lower 900 feet of the volcano is made up of fragments of pillow basalt, a bulbous, lumpy form of lava normally created during underwater eruptions. Overlaying this is 500 feet of volcanic clinker deposits, capped by 200 feet of basalt lava flows, both deposited once the vent surfaced above the glacier or lake.

For a fascinating close look at some of the layer-cake remnants of these eruptions, 0.6 mile north of the junction of FR 65 and FR 30, head west on

Aptly-named Lone Butte, seen here from Steamboat Mountain, is an example of a volcano that initially erupted beneath either glacial ice or a meltwater lake.

FR 3000401, a steep, single-lane, gravel track. To save your car, you may wish to park at the junction and hike the short distance. In 0.6 mile, the jeep road leads to a quarry on the southwest side of Lone Butte. Here the scooped-out hillside is clearly divided into two segments. The lower talus slope consists of fragments of pillow lava that erupted underwater; it is neatly overlaid by basalt flows from the vent after it broke through the glacial or water surface.

6

Mount Adams South

In volume, Mount Adams is one of the largest of the younger stratovolcanoes that line the Cascade crest; it is surpassed only by Mount Shasta. Adams is made up of nearly 48 cubic miles of accumulated andesite and dacite lavas, and the lava field in which it stands covers an area of more than 480 square miles. Just as the string of volcano vents and cinder cones along the north–south axis of the Indian Heaven lie along a zone of weakness, or fracture, in the earth's underlying crust, so too, more than sixty volcano vents are found along the axis of a major 35-mile-long fault that runs northwest–southeast through Adams and King Mountain, 11 miles to its southeast.

The earliest activity from the Mount Adams/King Mountain Fissure Zone occurred around 940,000 years ago when a vent, now buried under Adams, erupted multiple flows of blocky olivine basalt. Little of this lava remains exposed today; most was overlaid by the andesite flows that began around 520,000 years ago and formed present-day Mount Adams, and by a massive flood of olivine basalt that later issued from the King Mountain shield volcano.

The original Mount Adams was centered on a vent about 3 miles southeast of the present summit, in the headwaters of Big Muddy Creek. Eruptions of andesite and dacite between 520,000 and 490,000 years ago built a volcano cone here that was nearly as large as the present-day mountain. Today the only traces of this volcano are the glacier-eroded core on the ridge immediately north of the Ridge of Wonders and a host of dikes on the north side of that ridge that radiate from the center of the old volcano. Other vents erupted between 475,000 and 385,000 years ago on the west flanks of the older volcano and on its east side near The Island on the Ridge of Wonders and at the head of Cougar Creek.

The eruptive center shifted northwest to the vicinity of the present summit around 460,000 years ago, and a succession of andesite flows built a broad apron around the new vent, except on the southeast side, where they were probably blocked by the earlier cone. The current Mount Adams volcano continued to grow over the succeeding 300,000 years with ongoing eruptions of andesite lava; other vents on the northeast side of the mountain, in the vicinity of Goat Butte, extruded extensive basalt flows about 160,000 years ago.

A surge of basaltic lavas created several shield volcanoes southeast of Mount Adams some 120,000 years ago; King Mountain was one of the

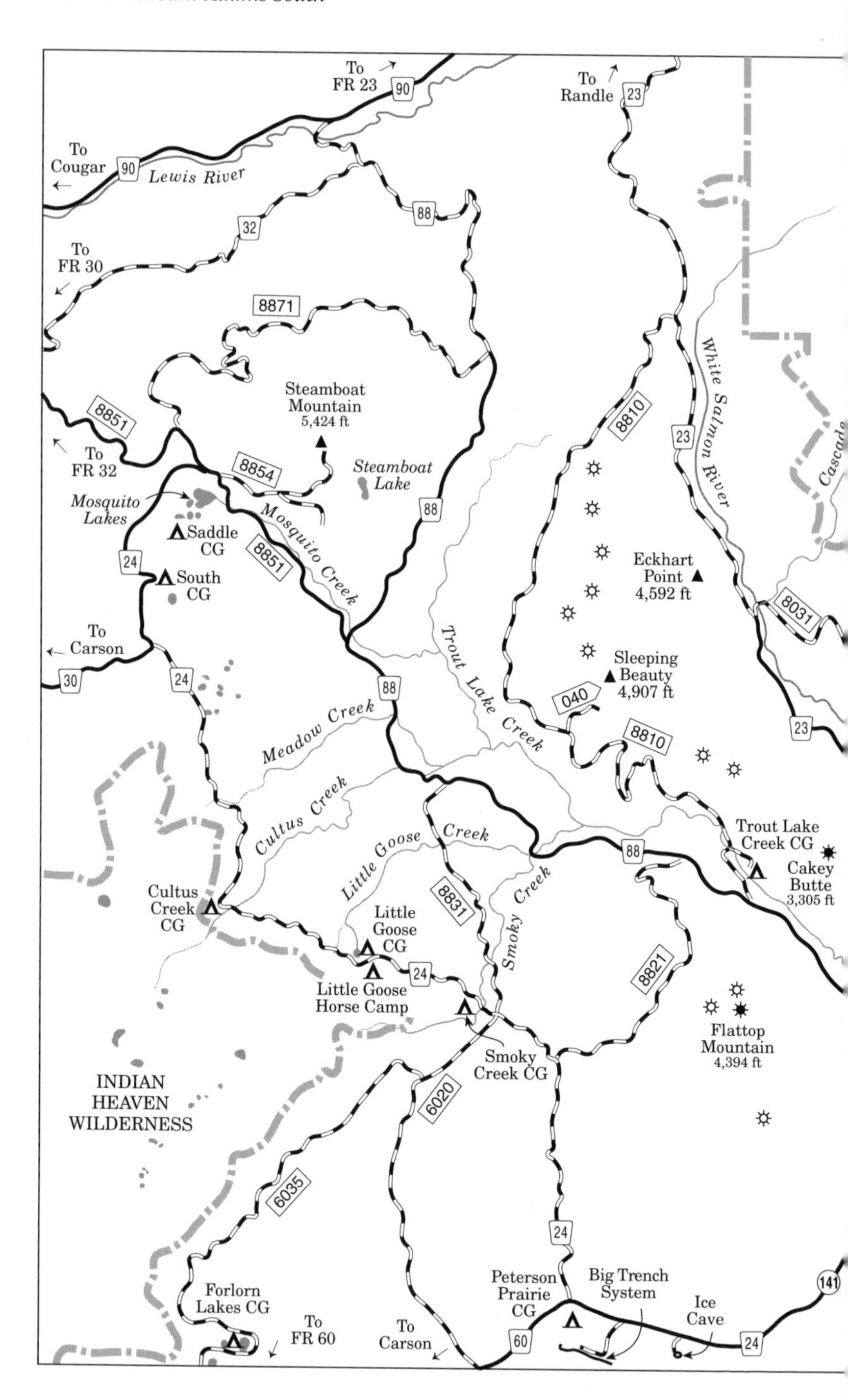
To FR 23
90
To Randle
23
To Cougar
90
Lewis River
88
32
To FR 30
8871
Steamboat Mountain 5,424 ft
8851
To FR 32
8854
Steamboat Lake
Mosquito Lakes
Saddle CG
Mosquito Creek
88
8810
23
White Salmon River
Cascade
24
South CG
8851
Eckhart Point 4,592 ft
8031
To Carson
30
24
Sleeping Beauty 4,907 ft
Trout Lake Creek
040
88
Meadow Creek
8810
23
Cultus Creek
Little Goose Creek
Trout Lake Creek CG
88
Cakey Butte 3,305 ft
Smoky Creek
Cultus Creek CG
8831
Little Goose CG
24
8821
Little Goose Horse Camp
Smoky Creek CG
Flattop Mountain 4,394 ft
INDIAN HEAVEN WILDERNESS
6020
6035
24
Peterson Prairie CG
Big Trench System
141
Ice Cave
Forlorn Lakes CG
To FR 60
To Carson
60
24

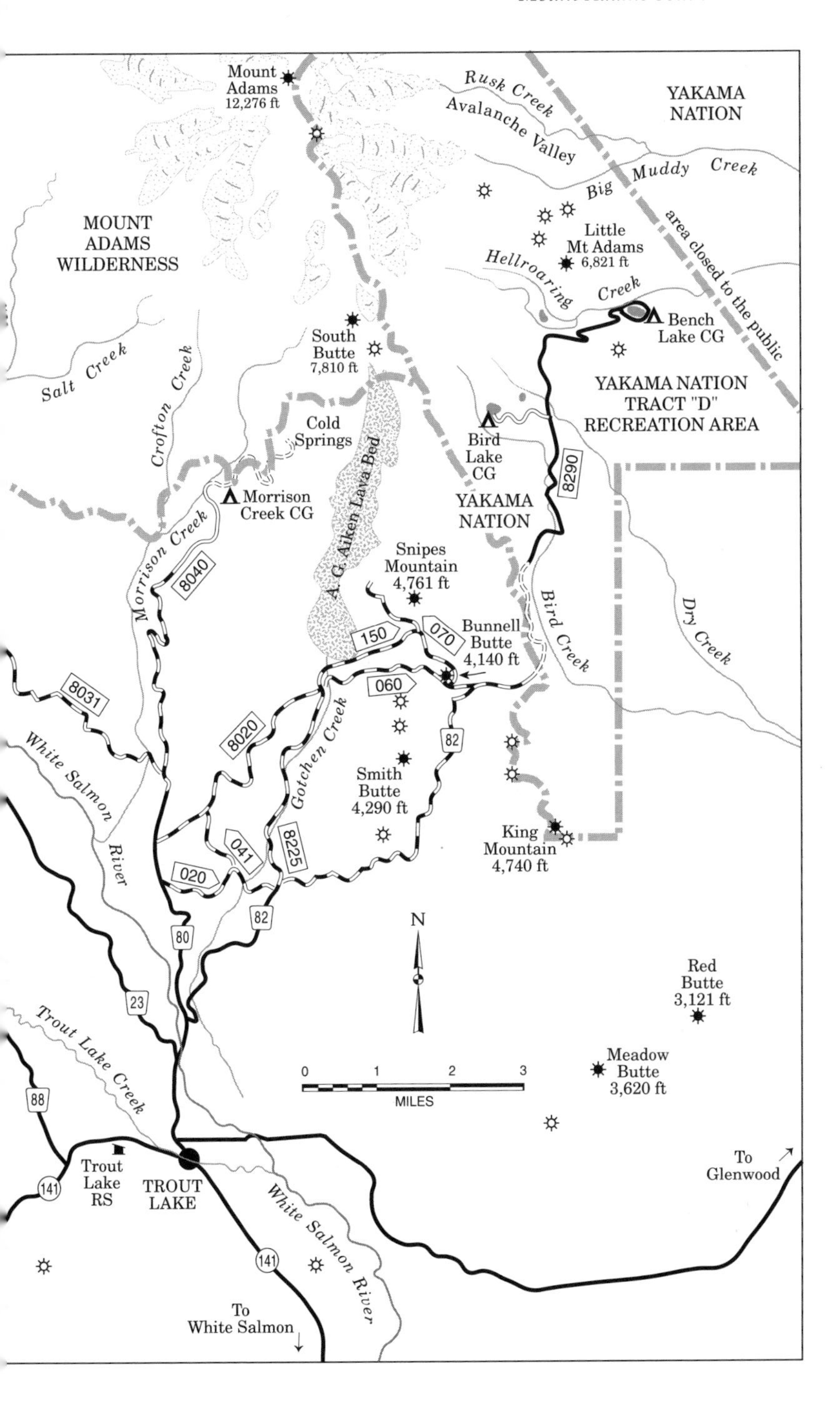
Mount Adams
12,276 ft
Rusk Creek
Avalanche Valley
YAKAMA NATION
Big Muddy Creek
area closed to the public
MOUNT ADAMS WILDERNESS
Little Mt Adams
6,821 ft
Hellroaring Creek
Bench Lake CG
South Butte
7,810 ft
Salt Creek
Crofton Creek
YAKAMA NATION TRACT "D" RECREATION AREA
Cold Springs
Bird Lake CG
8290
A. G. Aiken Lava Bed
YAKAMA NATION
Morrison Creek CG
Morrison Creek
8040
Snipes Mountain
4,761 ft
Bird Creek
Dry Creek
Bunnell Butte
4,140 ft
150
070
060
8031
8020
Gotchen Creek
82
White Salmon River
Smith Butte
4,290 ft
King Mountain
4,740 ft
041
8225
020
82
80
N
Red Butte
3,121 ft
23
Trout Lake Creek
0
1
2
3
MILES
Meadow Butte
3,620 ft
88
To Glenwood
Trout Lake RS
TROUT LAKE
141
White Salmon River
141
To White Salmon

largest. About 9,000 years later Potato Hill, on the fault line north of Adams, erupted a flood of basalt and formed a large cinder cone. Volcanic activity suddenly ground to a halt about 110,000 years ago, and little occurred over the next 30,000 years.

Volcanism resumed about 68,000 years ago with basalt eruptions northeast of Adams at Glaciate Butte, followed shortly by those from the cinder cone of Little Mount Adams. Several 20,000- to 12,000-year-old vents, topped by cinder cones, erupted multiple, thin layers of basalt lava that accumulated to depths of up to 150 feet between Mount Adams and King Mountain. Among these were Snipes Mountain, Bunnell Butte, Smith Butte, and several nearby unnamed cinder cones. The andesite flows that built the current summit of the mountain continued, with the present craters evolving around 15,000 years ago. Scattered around the base of Mount Adams are a handful of younger vents that poured out small flows of blocky andesite between 6,000 and 3,000 years ago.

Two main driving routes access the south side of Adams: one heads up the Little White Salmon River along the east side of Big Lava Bed; the second goes up the White Salmon River to Trout Lake. Both are paved roads leading north from Highway 14. The area can also be approached from the north by more circuitous and, most often, single-lane gravel roads.

Little Huckleberry Mountain

Hiking trail to the top of a dacite intrusion

Trail: Little Huckleberry Trail #49
Rating: (D); hikers, mountain bicycles, saddle and pack stock
Distance: 2.5 miles
Elevation: Trailhead 2,995 feet, Little Huckleberry summit 4,781 feet
Maps: USGS Little Huckleberry Mountain, USFS Mount Adams Ranger District
Driving Directions: Turn north from Highway 14, 12 miles east of Stevenson (9.2 miles west of White Salmon) on the county road signed to the Little White Salmon Fish Hatchery and Little White Salmon Recreation Area (2-lane paved). Farther north the road becomes FR 18 (2-lane paved). In 8.2 miles head west, then north on FR 66 (2-lane paved for 10 miles, then 1-lane gravel) for 12.2 miles. Trail #49 starts from the junction of FR 66 and FR 6600080.

A real tongue-dragging hike that gains nearly 1,000 feet of elevation per mile leads to a one-time fire lookout site. The original cupola cabin, built in 1924, was replaced in the 1930s by a standard lookout tower. The tower was abandoned in 1970 and was destroyed; all that remains atop the mountain are a few cement blocks, some melted glass, a tiny storage shed under a tree, and incomparable views.

The trail begins steep and maintains the relentless grade upward, along the west side of a stream drainage, climbing through a forest of mountain

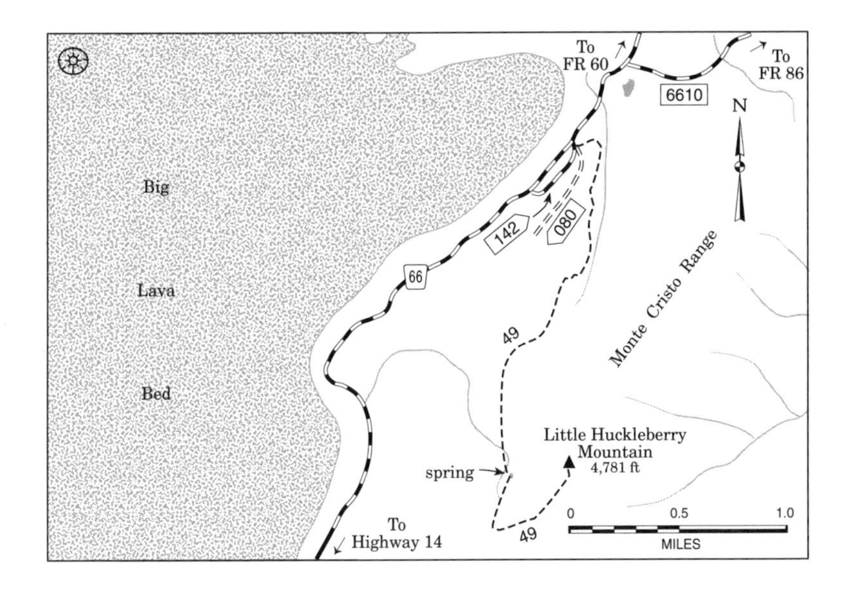

hemlock and fir. At 4,000 feet, after a well-deserved pause in the climb, the path does the unconscionable—it loses 100 feet in a long, sidehill traverse. In 0.2 mile the way turns uphill again, passes a pair of intermittent springs, then makes a long switchback and heads northeast toward the summit.

Halfway up the remaining slope the route breaks from the trees into a ridge-top rock garden. In midsummer rainbow hues of columbine, tiger lilies, Indian paintbrush, scarlet gilia, and lupine decorate the path to the summit. To the south, the valley of the Little White Salmon forms a foreground frame for icy Mount Hood. Northeast, Mount Adams slumbers on the skyline.

In the long plain to the west, gray-green forest cover marks Big Lava Bed; it sharply contrasts with the deeper green of the timber that wraps around the perimeter of the bed. Surprisingly, this deeper green forest also covers the cone from which the flow originated. Northwest is the backbone ridge of the Indian Heaven Wilderness, broken by slabby core remnants of old volcanoes.

Below, to the north, a patchwork-quilt of clearcuts spreads across nearby South Prairie and middistant Peterson Prairie. South Prairie is a chain of marshes and intermittent lakes, formed when their normal drainage to the southwest was blocked by Big Lava Bed flows. Peterson Prairie is on the surface of a series of basalt flows from the Lake Wapiki crater that surged east through the valley between 18,000 and 12,000 years ago.

Little Huckleberry Mountain, in contrast to many of the surrounding cinder cones, was not created by a volcanic eruption. It, and much of the ridge to the northeast, was formed about 20 million years ago by a non-eruptive intrusion of dacite through older volcanic flows. Subsequent erosion has exposed the hard, weather-resistant heart of this time-frozen magma.

The Big Trench System (Natural Bridges)

Short footpaths around the perimeter of the trench system

Rating: (E); hikers
Elevation: 3,000 feet
Maps: USGS Little Huckleberry Mountain, USFS Mount Adams Ranger District
Driving Directions: From Trout Lake, continue west on Highway 141 for another 6 miles, where it becomes FR 24 (2-lane paved). In 1.8 more miles turn south on FR 2400041 (1-lane gravel), and in 0.5 mile turn west on FR 2400050 (1-lane gravel). Park beside the road in about 0.3 mile.

In the heart of Peterson Prairie, west of Trout Lake, two natural bridges span a long, sinuous, 20- to 30-foot-deep trench that parallels the south side of FR 2400050. The bridges are remnants of the roof of an old lava tube called the Big Trench System. This was once an underground maze of basalt-flow tunnels nearly 4,000 feet long. Large portions of the tube have collapsed since it was formed some 12,000 years ago, and today the system consists of two natural bridges and six small caves, connected by open sections of the tube where the roof has collapsed to form a wide trench.

The caves are woven into a Klickitat Indian legend of infidelity and murderous revenge. A man and his wife once lived on the Columbia River near here. The man fell in love with another woman named Mouse and ran away with her to live in a hole in the ground near this area. The wife followed them, and in their flight from her they ran into an underground passage that emerged in a lake. The husband saw his wife and begged her to kill only her rival, but not him (the cad!). After killing the other woman, the wife became consumed with rage and also killed her husband. She then withdrew into a deep hole in this cave system, where she remains to this day. The Klickitats call the trench Hool Hool Se, derived from Hool, their word for mouse.

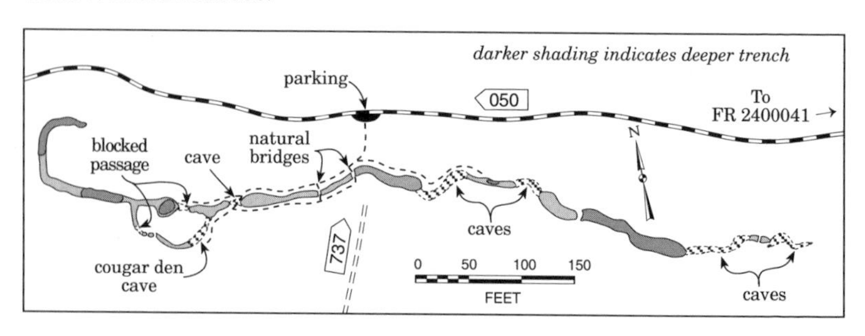

This natural bridge was formerly a portion of a lava tube that was formed 12,000 years ago.

A well-trod dirt path leads 100 feet south from the parking area on FR 2400050 to the edge of the section of the Big Trench that includes the two natural bridges. Unfortunately, a thick growth of thimbleberry bushes and other brush covers most of the floor and walls of the 200-foot-wide trench, obscuring many features of the former tube. At the east end of this section of trench is the wide, black mouth of one of the bigger caves in the system. The cave, which can be reached by scrambling down the steep rock wall along its edge, is about 200 feet long and 10 to 20 feet high. The floor is a rough jumble of basalt boulders. An interesting feature just inside the entrance is a pit that drops to a lower-level tube that is largely filled with breakdown debris, indicating that the original tube had a complex, multilayered structure.

The system continues east for another 1,500 feet or more, where three more caves are connected by more trench sections, one up to 75 feet deep. Another shallower one has an inner sink in its floor, probably also a remnant of a lower level of the initial tube. At the west end of the Natural Bridges sink is a smaller cave, easier to reach, and only about 4 to 6 feet high and 35 feet long. Farther west the trench hooks north, alternating between 40-foot-deep sections and shallower, 20-foot-deep ones, possibly more examples of an overlapping tube structure. This portion of the trench also shows evidence of a reentering side tube to the south, marked by shallow sinks and another 130-foot-long cave, once said to have been used as a cougar den. Most of the caves have a rich variety of plants, molds, and other biota.

Lava Tubes of the Ice Caves Basalt Flows

Between 18,000 and 12,000 years ago, the long valley that leads east from the Indian Heaven Wilderness to Trout Lake, and then south toward Husum, was flooded by lava. The source of the lava was a volcano vent in the heart of the crater that is now partially filled by Lake Wapiki. Several layers of hot, fluid basalt surged across the area, accumulating to a thickness of up to 100 feet. During these flows subsurface lava tubes formed, in the same manner as those in the cave basalts on the south side of Mount St. Helens, described in chapter 2. The Dry Creek, Peterson Prairie, and Trout Lake areas are laced with complex lava caves and trenches formed from collapsed tubes. In this area at least two-dozen lava tube caves are known and have been explored; the Forest Service is presently working on a more complete inventory of the tubes and their significant geological and biological features. Many of the tubes lie outside the National Forest boundary, on private property. One of these caves, no longer open to the public, is called Cheese Cave. The temperature and humidity were found to be identical to cheese caves in Roquefort, France, and for some years it was used to age Mount Adams brand cheese.

These caves contain a diversity of rare biota and lava formations that are easily damaged or destroyed, even by the most cautious visitors. As a consequence, the Forest Service discourages their exploration and, with a few exceptions, will not provide information on their locations. The authors respect the need to protect these natural treasures; we describe here only the two where the Forest Service provides public access.

Ice Cave

Scramble in a lava tube cave

Elevation: 2,820 feet

Maps: USGS Little Huckleberry Mountain, USFS Mount Adams Ranger District

Driving Directions: From Trout Lake continue west on Highway 141 for another 6 miles, where it becomes FR 24 (2-lane paved). In 1 more mile turn south into Ice Cave Picnic Area (day-use only).

This 650-foot-long cave was known well before the turn of the century. In a book published in the late 1860s, R. W. Raymond described a trip into the ice caves of Washington Territory. He noted that Portland hotels and taverns were forced to make do with disgusting tepid drinks in summer; however, at Dalles City (the Dalles) bars offered civilized ice-mixed drinks. In pursuit of the source of this extravagance, Raymond took a steamer up the Columbia to the mouth of the White Salmon River, then set out with a party on horseback to reach the source of the ice.

Following Indian trails up the White Salmon, the author related encounters with native berrypickers, swarms of yellowjackets, and mule trains packing blocks of ice downstream to the Columbia River. After 35 miles of cross-country travel the group arrived at the ice caves. Raymond described his exploration of this natural wonder in the colorful, flamboyant style typical of travel writers of the era.

Today's Ice Cave is somewhat the worse for wear, with ice deposits and crystal stalagmites and stalactites diminished, possibly due to recent warmer-than-normal winters. Nonetheless, slippery floors of ice persist in the lower east end of the cave, where cold air and winter snows are trapped, and temperatures remain near or below freezing year-round.

Don't bother with Raymond's cross-country route to reach the caves. Instead, drive FR 24 to the day-use picnic area. East of the parking lot is a 14-foot-deep sinkhole entrance to the cave where the Forest Service has an interpretive display and has built a staircase that descends to the base of the pit. Safe exploration of the subterranean chambers demands good footwear, head protection, warm clothing, and two or more reliable sources of light.

The floor is covered with blocky lava debris and slick ice. To the east, the cave descends 120 feet to a tube-end ice pool and the Crystal Grotto, which, after a winter of low temperatures and heavy snowfall, becomes a fantasy land of floor-to-ceiling ice columns, glass-like domes, and hanging crystal blades.

West from the base of the stairs, the ice-glazed floor leads past a peephole that looks into a low, arched chamber in the north wall. The path continues past lava flow imprints in the floor to a small lava bridge in front of a diminutive inner chamber lined with ice crystals. To the west a large, foreboding pile of basalt debris constricts the tube; beyond, the cave twists westward to a pit and debris-fall that blocks further exploration.

Three other sections of the cave to the west are accessible from two more intervening sinkholes. To reach the wide, low, westernmost cave requires an uncomfortable crawl through a rough, constricted channel. The

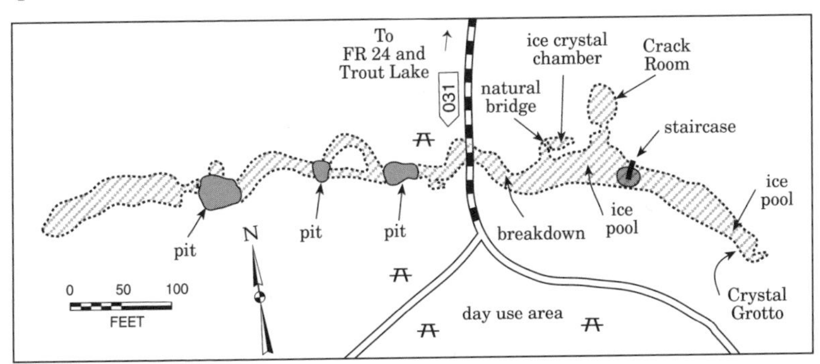

Ice Cave is a lava tube that is so cold that dripping water forms into a fantasy of ice columns.

middle cave section is about 100 feet long and 15 feet high, with debris breakdown on the floor offering the only obstruction. The easternmost section, which probably connected to the main portion of the cave at one time, is made up of two smaller tubes that split, then rejoin. Both start stoop-shoulder-high and taper to crawl-size as they approach the piles of debris that block the west end of the main cavern.

Sleeping Beauty

Hiking trail to the top of an andesite intrusion

Trail: Sleeping Beauty Trail #37
Rating: (M); hikers
Distance: 1.4 miles
Elevation: Trailhead 3,500 feet, Sleeping Beauty summit 4,907 feet
Maps: USGS Sleeping Beauty, USFS Mount Adams Ranger District
Driving Directions: 1.7 miles west of Trout Lake, turn north on FR 88 (2-lane paved), follow it for 4.8 miles, then head northeast on FR 8810 (1-lane gravel). In 5.3 miles turn east on FR 8810040 (1-lane gravel), and in another 1.4 miles arrive at the start of Trail #37.

One of the most striking landmarks on the west side of Mount Adams is the vertical, 500-foot-high south face of Sleeping Beauty. According to

Indian legends this is the beautiful maiden that captured the hearts of both Wy-east (Mount Hood) and Pah-toe (Mount Adams) and caused the two brothers to fight violently (see Bridge of the Gods in chapter 3). As punishment for her trouble-making, the Great Spirit changed her into this peak, whose summit, some say, resembles the profile of a sleeping woman, lying on her back. When approaching Trout Lake from the south, the peak is the visual capstone of the Trout Lake Creek drainage. From the southwest, the sheer south wall of the peak is a "must-include" foreground in photographs of the glacier-clad flank of Adams.

The narrow, rugged summit of Sleeping Beauty is a core of andesitic magma that intruded up into older volcanic strata about 25 million years ago. This hard, rock column was exposed as the soft older layers around it eroded away. From the summit, columnar basalt flows and outcrops can be seen on the lower logged-off ridge to the north. These are remnants of lava that erupted from six small volcano vents along the ridge about 570,000 years ago.

Trail #37 leaves FR 8100040 and immediately starts to climb through dense undergrowth and second-growth forest. The tread is good and well

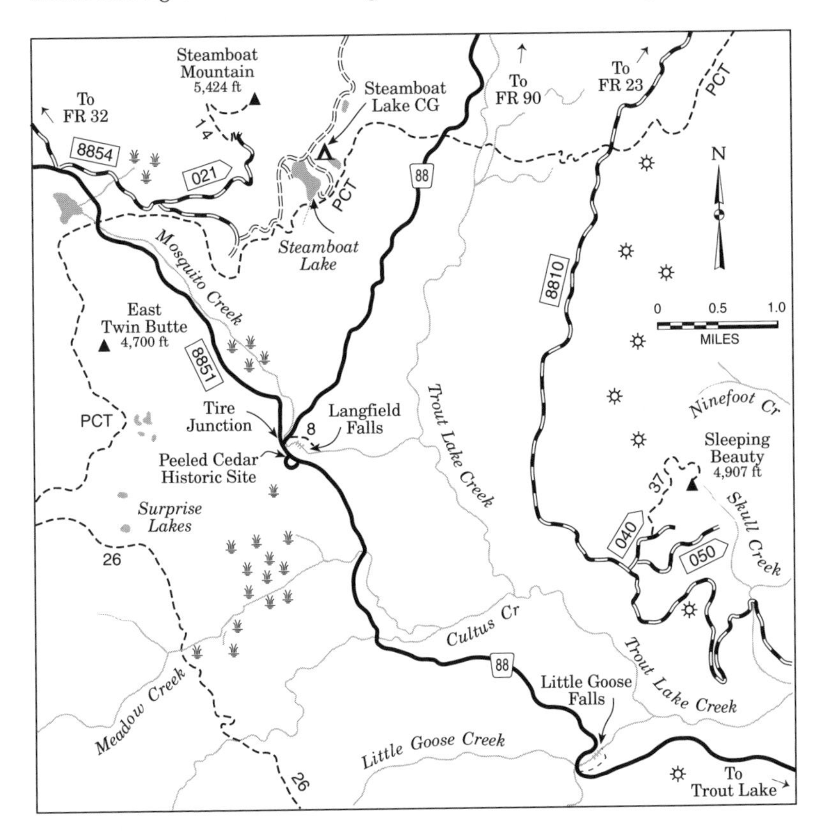

Mount Adams, a stratovolcano, watches over the rugged summit of Sleeping Beauty.

maintained, but the grade is unrelentingly steep! In about a mile the route approaches the ridge top, the uphill grind eases, the undergrowth fades away, and the forest changes to old-growth mountain hemlock and Douglas-fir.

The summit cliffs to the southeast, although only 200 feet high on the north side, look impenetrable. As the trail reaches the base of this outcrop wall, it snuggles next to the rock and winds up through narrow cracks over a tread built from stacked layers of rock. The path resembles one through ramparts to the castle of Oz's Wicked Witch of the West—take care my pretty one!

The rugged summit crest is about 750 feet long but only 15 to 30 feet wide, with precipitous dropoffs on all sides. Eye bolts set into its surface are from a fire lookout that sat atop the peak between 1931 and the 1960s. The views are breathtaking! To the northeast, only low wooded ridges and clearcuts stand between the peak and the horizon-filling mass of Mount Adams. Long valleys stretch south down the White Salmon River drainage to Trout Lake and down the prairie floor east of the Indian Heaven Wilderness to Big Lava Bed. On the skyline beyond is sharp, icy Mount Hood. The entire Cascade crest running through the Indian Heaven Wilderness is in view, topped by Lemei Rock and Red, Bird, and Sawtooth mountains.

Little Goose Falls

Cross-country scramble to a waterfall

Rating: (M); hikers
Elevation: 3,080 feet
Maps: USGS Sleeping Beauty, USFS Mount Adams Ranger District
Driving Directions: From Trout Lake, take Highway 141 west for 1 mile, then turn northwest on FR 88 (2-lane paved), and in 8.2 miles park at a small pulloff on the east side of the road, just before crossing the bridge over Little Goose Creek.

Here is an opportunity to examine three different geological phenomena and the impact of erosion on each, to enjoy a secluded picnic spot bedside a pretty little waterfall, and to stuff yourself with huckleberries en route. From the narrow pulloff on the south side of the bridge over Little Goose Creek, follow a boot-beaten path east (downstream) through trees, along the high bank above the creek. In about 100 yards the trees give way to a large clearcut covered with huckleberry bushes that are heavy with tasty fruit in early fall.

Carefully approach the canyon wall above the creek; it drops precipitously for 100 feet to the creek bank. Glimpses upstream through cliff-face trees disclose the white ribbon of a waterfall but reveal no feasible way to safely get down to it. Continue east along the canyon rim for 200 yards to where the boot path finds a chink in the canyon wall and, with the assistance of brush and tree root handholds, slither down to the creek bed. Head back upstream to the waterfall and discover—a picnic table! How it got here is a mystery. Don't question, just enjoy.

The waterfall splits around the sides of mid-lip lava rock then drops about 25 feet, like a frothy beard, into a plunge pool so dark that it appears bottomless; surely it must harbor some fantasy monster in its depths! Now for the brief geology lesson.

The light gray, blocky, 150-foot-high wall on the opposite side of the creek is olivine basalt, erupted from small volcano vents about 3 miles to the northeast more than 140,000 years ago. At the falls this flow butts against a 20,000-year-old flow of basalt from the crater to the southwest, presently occupied by Lake Wapiki. This lava was subsequently overlaid by glacial till, deposits from melting alpine glaciers about 11,000 years ago. Little Goose Creek has since cut through the softer glacial deposits, as well as older formations beneath the two lava flows, exposing the basalt wall on the north side of the creek and creating the waterfall as it wore through the glacial till to the harder underlying lava flow that now forms the lip of the falls.

Langfield Falls

Footpath to a waterfall

Rating: (E); hikers
Elevation: Trailhead 3,480 feet, base of falls 3,400 feet
Maps: USGS Sleeping Beauty, USFS Mount Adams Ranger District
Driving Directions: From Trout Lake take Highway 141 west for 1 mile, then turn northwest on FR 88 (2-lane paved), and in 14.8 miles reach Tire Junction—the source of the name is obvious. In 0.1 mile to the northeast pull into the parking lot at the head of the Langfield Falls Trail #8.

Langfield Falls streams over a layer of erosion-resistant basalt.

An easy trail wanders downhill to one of the most beautiful waterfalls in the South Cascades. The broad, gentle-grade path descends along the forested hillside with ever-enticing views of this gorgeous cascade. At a switchback below the falls, an interpretive sign tells of the lifetime dedication to the Gifford Pinchot Forest by Casey Langfield, for whom the falls is named. Langfield was District Ranger of the Mount Adams District between 1933 and 1956.

Below, the path leads back to the base of the falls, a broad 200-foot-high film of water that splashes off basalt nodules protruding from the wall beneath in a endless series of miniature, V-shaped veils.

The falls were created by Mosquito Creek as it raced downhill, cutting through overlying layers of local alpine glacial deposits, before reaching harder underlying basalt. The basalt flows were laid down about 275,000 years ago during eruptions from the Sawtooth Mountain volcano. The black nodules in the face of the falls, surface examples of pahoehoe lava flows, are mute reminders of the volcanic history of the area.

Steamboat Mountain

Hiking trail to the top of lava flows

Trail: Steamboat Mountain Trail #14
Rating: (M); hikers
Distance: 1 mile
Elevation: Trailhead 4,840 feet, summit 5,424 feet
Maps: USGS Steamboat Mountain, USFS Mount Adams Ranger District
Driving Directions: Follow directions to Tire Junction as described in Langfield Falls, preceding, then continue north on FR 8851 (1-lane paved with turnouts). In 3.2 miles head east on FR 8854 (1-lane gravel), and at a Y-junction in another 0.8 mile continue east (left) on FR 8854021 (1-lane gravel). This road ends in 1.6 miles at a gravel pit on the southeast face of Steamboat Mountain, the start of Trail #14.

This old lookout site offers wonderful views of Mount Adams, the Goat Rocks, the crest of the Indian Heaven Wilderness, and the backbone ridges of the Dark Divide. However, access may be closed if the Forest Service decides that recreational use is not compatible with the Natural Research Area encompassing the surrounding forest. Perhaps a little friendly lobbying might be in order to prove that such a closure is unwarranted.

For the present anyway, the trail starts at the edge of a huge gravel pit (Natural Research Area compatible?) with an outstanding view of the 300- to 500-foot-high wall of volcanic strata that stretches northeast for nearly ½ mile. This is an impressive exposure of three thick flows of Grande Ronde basalt, the only occurrence of this massive lava flow on the west side of Mount Adams. Between 17 and 15.5 million years ago the lava spread across the Columbia Basin from the Idaho–Oregon–Washington border to the mouth of the Columbia River. Grande Ronde basalt was originally deposited on lowlands near sea level; its occurrence above 4,500 feet provides evidence of the extent of uplift in the Cascades over the past 15 million years.

The trail climbs steadily up the southwest flank of the mountain, through western hemlock forest with a thick understory of huckleberries and random bursts of hellebore. In 0.5 mile the path reaches a saddle northwest of the summit, then turns up an open grassy ridge in search of the mountain top. The trailside is a veritable garden of wildflowers: paintbrush, stonecrop, self-heal, penstemon, and yarrow. One of the authors found trees along the path swarming with Oregon juncos, always one jerky, limb-to-limb flight ahead of the hiker.

At the summit, just a step back from the vertical, southeast face, is a square of concrete blocks, part of a fire lookout that stood here between 1927 and the early 1960s. The choice of this location is obvious—the view east is of forest lands all the way to Mount Adams; west, the truncated summit of St. Helens; northeast, the backbone of the Goat Rocks lining the horizon; north, the rocky summits along the fingers of Juniper Ridge; and southwest, the crest of the Indian Heaven Wilderness, the cinder cones of Twin Buttes, and the isolated lava pyramid of Lone Butte.

Mount Adams Summit

Climb to the summit of a stratovolcano

Trail: South Climb Trail #183
Rating: (D); hikers, saddle and pack stock
Distance: 6.2 miles
Elevation: Cold Springs Trailhead 5,590 feet, end of #183 8,000 feet, summit 12,276 feet
Maps: USGS Mount Adams East, USFS Mount Adams Ranger District
Driving Directions: Head north from Trout Lake on Highway 17, signed to Glenwood. At a junction in 2 miles take the left fork, FR 80, signed "South Climb Trail #183." The road is 1-lane paved with turnouts for the next 3.8 miles to the junction with FR 8040, where it becomes teeth-jarring washboard gravel for 3.6 miles. Beyond, it is signed as narrow, rough, and single-lane. Despite that warning, continue north for another 2.1 miles to Morrison Creek Campground. Here the road becomes FR 8040500 (1-lane dirt), deteriorates dramatically, and in 2.9 miles reaches its end at Cold Springs Trailhead.

The south side route up Mount Adams via Suksdorf Ridge is not a technically challenging climb—in the early 1930s the route was used to supply the summit lookout, and when that was abandoned the trail was maintained for teams of pack mules to haul sulfur mined from the summit crater. However, this does *not* mean that this climb is a Sunday stroll with no hazards; there have been serious injuries, and even deaths, on the route. Climbers typically take from six to eight hours to ascend nearly 6,700 vertical feet across snow and rock to reach the 12,276-foot summit. This is a physically draining effort, and the rapid change in elevation can trigger mild-to-severe altitude sickness. Bad weather and poor visibility can be a major hazard; even on a sunny summer day local cloud banks can quickly envelope portions of the route, turning everything into a dimensionless white soup. The climb is on pumice and snowfields; however, crevassed glaciers on either side are hazards for those who stray from the route.

Climbers must register at the ranger station at Trout Lake before starting up the mountain and check out upon their return. In addition to the Ten Essentials, ice axes and crampons (and knowledge of their use) and ample warm clothing are advised. If there is any possibility of weather closing in, marking the portion of the route above the end of the trail with wands is a sensible precaution. Remove wands upon descending.

From Cold Springs Trailhead, Trail #183 follows an abandoned jeep track for 2 miles to primitive Timberline Camp at 6,260 feet (no water). From here the route winds briefly upward through stubby alpine fir to barren slopes that are either snow-covered or pumice, depending on the season. The trail continues uphill to the north, bearing slightly west to the rib along the west side of the Crescent Glacier (snowfield). The trail is maintained to about 8,000 feet, and a boot path can be followed to about 9,000 feet, where there is a level rest spot known as the Lunch Counter. The route above bears east to the top of Suksdorf Ridge and follows it north

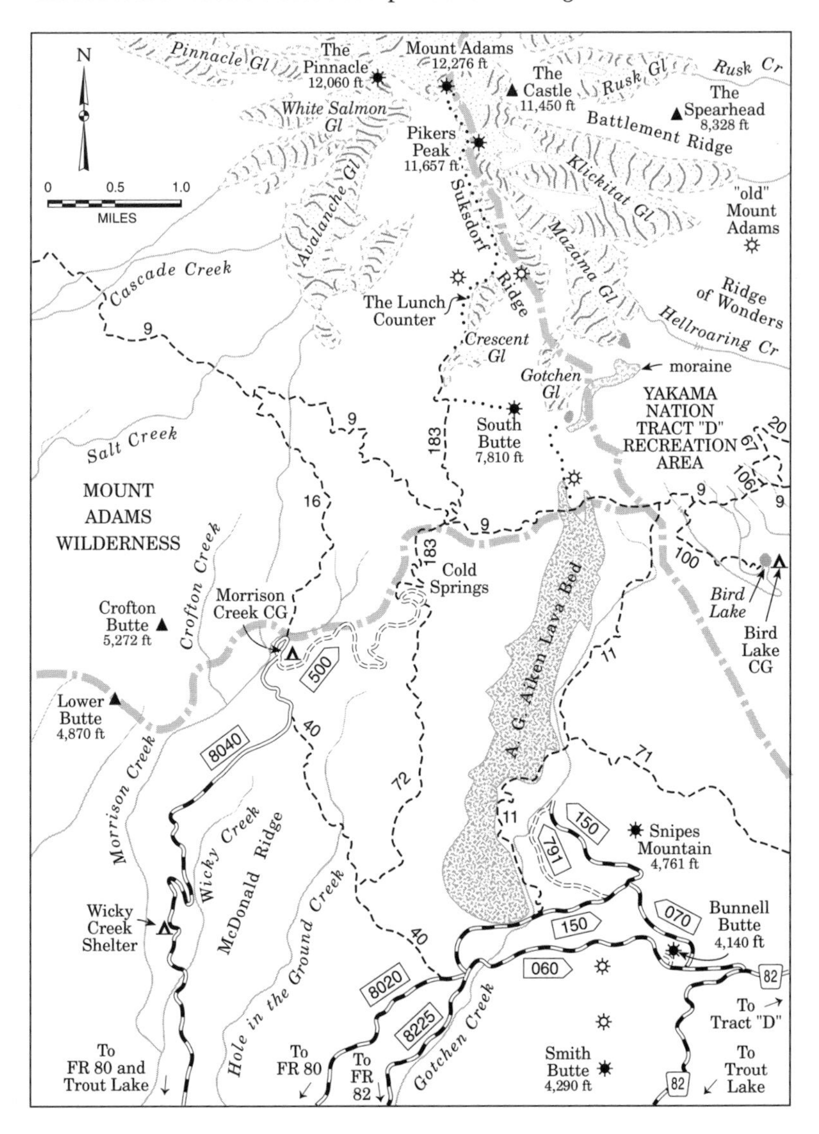

A climbing route via Suksdorf Ridge leads to the summit of the mighty stratovolcano of Mount Adams.

to the false summit, Pikers Peak, at 11,600 feet. The climb then heads northwest to the summit crater; the true summit is on its north rim.

Between 1918 and 1921 the Forest Service built a fire lookout atop the peak but soon found that supplying the lookout was a major effort and frequent cloud cover over the upper slopes of the mountain limited its usefulness. The lookout was abandoned in 1924, but the cabin was modified in 1932 for use by miners extracting sulfur from the crater floor. Although the mine shipped over 180 tons of sulfur out via pack mules, by 1937 new sulfur extraction techniques at more accessible sites made the operation obsolete, and all active mining ceased. The summit cabin is jammed with ice and buried beneath a blanket of snow but can still be found when portions of it are uncovered in late summer.

Although Adams is classified as one of the Cascade stratovolcanoes, like Rainier, Baker, St. Helens, and Glacier Peak, it has its own unique geological features. One becomes apparent when you observe the mountain from a northwest–southeast versus an east–west axis. From the northwest and southeast the volcano appears reasonably symmetrical, like the rest of the glaciated giants; however, when viewed from the east or west, its summit is much longer and broader than the others. It appears that the mountain top is made up not just of one primary vent but of three individual ones, or else over time there was a significant southeast shift in the location of the primary vent. The oldest of the three is on the northwest at the Pinnacle, the high point between the White Salmon and Adams glaciers. This is either a distinct vent or a portion of the rim of a much larger crater that was destroyed during eruptions that built a new crater to the southeast about 15,000 years ago. This new crater is the highest spot on the peak today. Farther southeast is the lower false summit, Pikers Peak, a separate, younger eruptive site around 13,000 years old, the source of the flows that make up Suksdorf Ridge.

Another peculiarity of Adams is that, unlike Rainier, it has no deep canyon-filling lava flows. The individual layers in its flows are all rather thin and uniform, indicating that eruptive cycles occurred with a frequency that did not allow enough time for deep canyons to be eroded into its sides between sequences of lava flows. Although the summit of Adams has shown no volcanic activity for over 10,000 years (other than occasional sulfurous fumaroles), several very young parasite cinder cones have erupted around its base within the last 4,000 to 2,000 years, extruding tongues of andesite and basalt that form blocky lava beds below the vents.

South Butte and the Gotchen Glacier Moraine

Hiking trail and cross-country route to a lava flow and glacial moraine

Trails: South Climb Trail #183, Around-the-Mountain Trail #9
Rating: #183 (D), #9 (M), cross-country (D); hikers, saddle and pack stock
Distance: Trailhead to cross-country start point 3.8 miles, cross-country 1 mile
Elevation: Trailhead 5,590 feet, cross-country start point 6,400 feet, Gotchen Glacier moraine 7,300 feet
Maps: USGS Mount Adams East, USFS Mount Adams Ranger District
Driving Directions: See directions to Mount Adams Summit, preceding.

Here is a good opportunity for close-up views of five distinct geological features on the south side of Mount Adams, with loads of scenery thrown in to boot. This adventure starts at the end of FR 8040500 at Cold Springs Trailhead, where South Climb Trail #183 heads toward the summit. In 2 miles, just at timberline, Trail #183 crosses Trail #9; head east here on Trail #9, which makes a twisting traverse around old lava ribs, sometimes in sparse low subalpine forest, other times through dry brown patches of pumice. In a little under a mile the trail crosses a black finger of blocky andesite near the top of A. G. Aiken Lava Bed, whose moonscape surface drops south for nearly 4 miles. Shortly the path crosses another smaller finger of the bed at about 6,400 feet, and here "the fun begins."

Leave the trail for a cross-country ascent north-northwest, heading toward the obvious rust-colored cone of South Butte. Pick your way uphill, following animal tracks through glorious flower-decked meadows, finding the easiest way up the steep slope. Meadows give way to boulder hopping, and then two-steps-forward, one-step-back travel over loose, ball-bearing-like tephra. At 6,700 feet pass a black, 85-foot-high cinder cone that sits atop the suspected source vent for A. G. Aiken Lava Bed.

At 7,400 feet bear northeast to the semicircular gravel rim below the Gotchen Glacier. About 100 feet below the rim is a small, oblong snowfield that, by early fall, transforms into a brilliant blue lake. The gravel rim around the south and west side of the lake is a terminal moraine, debris deposited at the lower end of a glacier. It marks the farthest advance of local alpine ice during a period of lowered temperatures that occurred between 4,000 and 2,000 years ago.

Nearby, to the west and 400 feet higher, is South Butte, a volcano vent on the lower end of Suksdorf Ridge. The south side of this cone is a mass of orange-red clinkers, while its north side is covered with dark black versions of the same. Lava cliffs between this point and the butte make cross-country travel to it dangerous. Those with an incurable desire to more closely inspect the top of the cone, where remnants of its lava core are displayed in three stubby basalt outcrops, should approach it from the summit climbing route to the west.

On the return trip downhill look out over the wooded hills to the south, and note that most have a steep-sided triangular shape. All are cinder cones atop volcano vents that flooded the terrain to the southwest with layers of olivine basalt over 20,000 years past.

The A. G. Aiken Lava Bed

Hiking trail beside a lava bed

Trail: Snipes Mountain Trail #11

Rating: (M); hikers, saddle and pack stock; mountain bicycles to the junction with #71

Distance: 5.8 miles

Elevation: Trailhead 3,800 feet, junction with #9 6,300 feet

Maps: USGS King Mountain, Mount Adams East; USFS Mount Adams Ranger District

Driving Directions: From Trout Lake head north on Highway 17. In 2.1 miles, at the junction with FR 80 (2-lane paved), bear east. In 2.2 miles the road becomes FR 82 (1½-lane gravel); in another 0.4 mile turn north on FR 8225 (1-lane gravel). At a four-way intersection in 2.2 miles, continue north on FR 8020150 (1-lane gravel) to Trail #11 in 0.8 mile.

This pleasant, wooded hike skirts the east side of A. G. Aiken Lava Bed, offering frequent close-up views of the bed from its base at the trailhead to where the trail turns to the east, about 3 miles farther and 1,400 feet of elevation higher. The Aiken bed was formed by the most recent eruption near Adams, thought to have happened between 4,000 and 2,000 years ago. Its surface is a massive jumble of dark gray andesite blocks, most watermelon- to desk-sized, randomly piled to a height of 50 to 130 feet above the surrounding landscape. The surface of the flow is a blocky lava moonscape with only a few trees and a sparse growth of moss and lichens, both testifying to its young age.

Trail #11 starts out through a forest of big Douglas-fir, hemlock, and ponderosa pine. The tread is good, and the grade is gentle, although persistently uphill. For those whose interests stray from geology to botany, trailside wildflowers range from yarrow and a few columbine to pockets of beargrass and sprinklings of scarlet gilia amid thick mats of purple asters.

Switchbacks ease the upward travel, and at the west end of each the track nudges close enough to the lava bed that the curious can readily climb up the rough boulder sides for better views of the 4-mile-long by ½-mile-wide tongue of lava blocks.

Trail #9 crosses the top of the A. G. Aiken Lava Bed.

At 2.8 miles the path meets Trail #71 from the east. This junction marks the end of permitted mountain bike travel from both directions; as if to emphasize the point, the path immediately steepens. The way bends east 0.2 mile farther and climbs away from the lava bed. The forest soon thins, shrinks to subalpine-sized stands, and breaks into small grassy flower meadows. Three miles beyond Trail #71 the way joins Trail #9 at the Yakama Reservation boundary. The start of the cross-country portion of the South Butte trip, preceding, is 0.5 mile west; 1 mile east is Bird Creek Meadows, described later.

Bunnell Butte

Road trip to a cinder cone

Elevation: 3,920 feet
Maps: USGS King Mountain, USFS Mount Adams Ranger District
Driving Directions: Head north from Trout Lake on Highway 17. In 2.1 miles, at the junction with FR 80, bear east. In 2.2 miles the road becomes FR 82 (1½-lane gravel). Continue northeast on FR 82 for another 5.4 miles to FR 8225060 (1-lane gravel), turn west on it, and in another 0.5 mile pull north into a gravel pit at spur FR 8225069 (1-lane gravel).

For a fascinating presentation of the structure of a volcanic cinder cone, take this short, ½-mile diversion while en route to Tract "D" (see the sidebar). The fault zone between King Mountain and Mount Adams is lined with a number of small cinder cones such as Snipes Mountain and Smith and Bunnell buttes, as well as other unnamed vents in the area. All contributed to a local field of basalt lava extruded about 20,000 years ago.

The Forest Service has hacked a gravel pit about halfway through the south half of Bunnell Butte, in the process creating a near-perfect

Tract "D"

The geological fault line between King Mountain and the Mount Adams summit also marks a fracture of another sort—political—between the United States government and the Yakama Indian Nation. Problems began in 1855 when Kamiakin, the great chief of the Yakama tribes and bands reluctantly signed the Yakima Treaty. This treaty, dictated by Governor Isaac Stevens, ceded 10,800,000 acres of Yakama lands west of the Cascade crest to the United States in return for the Indians' permanent ownership of 1,200,000 acres of the Yakama Reservation, which was to include Mount Adams, a peak held sacred by the tribe. The treaty identified, in words, the boundaries of the reservation, and it was accompanied by a map showing these boundaries.

The treaty was ratified in 1859, but its map became "lost" in government files. In 1897 President Grover Cleveland issued a proclamation creating the Mount Rainier Forest Reserve from all forested public lands in the area west of the Yakama Reservation. In 1906 the first survey of the reservation boundary established the Campbell Line, which officially defined its western boundary, as interpreted from the *words* of the treaty. The survey excluded lands to the west, including Adams, that the Yakamas claimed were part of the reservation as shown on the missing map. The tribe protested that this line so poorly reflected the terms of the treaty that it didn't even include the sacred Mount Adams—the closest it came to the summit was 3 miles to its east! However, a powerful lobby of settlers in the disputed area overpowered Yakama objections (which couldn't be substantiated because of the missing treaty map). The forest reserve was expanded in 1908 to include 21,000 acres of land east of Adams up to the Campbell Line, while the remainder of the disputed land was claimed or bought by settlers.

In 1923 the boundary was resurveyed, establishing the new Pecore Line, which, while adding more land to the reservation, still fell short of the treaty commitments, in the view of the Yakama tribe. They contended that it excluded 121,465 acres of land east and southeast of Mount Adams that belonged in the reservation. The original treaty cross-section of this volcanic feature. Layers of stark black, foamy pumice, occasionally interleaved with orange-colored clinkers, taper gradually up to the top of the cone. At its heart is a V-shaped wedge of denser basalt cinders oxidized to a rich rusty red. The picture is so graphic you can almost feel the ground shake and imagine hot popcorn-like fragments of basalt being ejected into the air around you.

map was rediscovered in the 1930s, but by that time 98,000 acres of the disputed territory had been sold to private landholders, and in 1942 the remaining national forest acreage was designated as the Mount Adams Wild Area. This land, plus another 35,000 acres to the west was incorporated in the Mount Adams Wilderness in 1964.

The Yakama Nation continued legal and political pressure to recover at least the 12,000 acres of disputed land that was now protected as a part of the wilderness. In 1972, President Richard Nixon signed an Executive Order transferring this land back to the Yakama Nation, with the understanding that it would continue to be administered as a wilderness and that existing public access and trails in the area (now known as Tract "D") would be maintained.

Tract "D" is open only between July 1 and October 1 and is subject to stringent regulations. Among these are the following: fees are charged for day-use as well as camping; special day, week, or season fishing permits are required; boats with gasoline motors, motorbikes, ATVs, fireworks, firearms, bows and arrows, slingshots, and B-B guns are all prohibited; natural features such as wildflowers, huckleberries, mushrooms, herbs, and shrubs may not be removed; and strict campground courtesy and cleanliness rules are enforced.

Campsites are provided at Mirror, Bird, and Bench lakes and Sunrise Point. All but Sunrise Point have fire grates, picnic tables, water, and pit toilets. No trailer hookups are provided (which is not a significant problem, because access roads are so bad that no one in their right mind would take a large trailer or RV over them).

To reach the recreation areas in Tract "D," head north from Trout Lake on Highway 17. In 2.1 miles, at the junction with FR 80, bear east. In 2.2 miles the road becomes FR 82, a 1½-lane gravel washboard. In 6 miles, at the Yakama Nation Boundary, the road becomes FR 8290 and in another 0.5 mile deteriorates to a very rough, pot-holed, rutted, narrow, single-lane gravel and dirt road. Mirror Lake and the turnoff to Bird Lake are reached in another 4 miles, the Bird Creek Meadows parking area in 5 miles, and Bench Lake in 6.5 miles.

Bird Creek Meadows and Hellroaring Creek View Loops

Hiking trails to views of glaciers, moraines, and lava flows

Trails: Bird Creek Meadows Trail #105, Flower Walk Trail #106, Crooked Creek Trail #100; Round-the-Mountain Trail #9, Trail #20, Trail #67
Rating: (M); hikers, #9 and #100 saddle and pack stock
Distance: Bird Creek Meadows Loop 4.3 miles, Hellroaring Creek View Loop 2.5 miles
Elevation: Bird Creek Campground trailhead 5,585 feet, Flower Walk trail high point 6,280 feet, Bird Creek Meadows trailhead 5,676 feet, Hellroaring Creek viewpoint 6,480 feet
Maps: USGS Mount Adams East, USFS Mount Adams Ranger District
Driving Directions: Follow the driving directions above to the Bird Creek Meadows parking area in Tract "D," preceding.

Meadows, wildflowers, waterfalls, lakes, glaciers, cliffs, moraines, cinder cones—name it and some combination of short trail loops in the Bird Creek Meadows area will deliver it—and more! The biggest challenge in the Bird Creek Meadows area is getting there; forest roads leading to the Yakama Nation boundary are spring- and shock-testing washboards, and once across the boundary the road deteriorates to a rutted, single-lane dirt road that is not the place to drive your mama's Maserati. However, once you arrive at Bird or Mirror Lake campgrounds, or the large Bird Creek Meadows parking area, campsites are clean and well maintained, and trails are in relatively good shape. Be aware that day-use as well as camping and fishing fees are charged by the Yakama Nation and will be collected by on-site rangers.

Two loop trails are described here, but another three or four can be put together from various trail segments that wind thorough Bird Creek Meadows. For the longest Bird Creek Meadow loop, drive west from FR 8290 at Mirror Lake for 1 mile to Bird Lake Campground. Here find Trail #105 at the north side of the lake, and follow it east for an easy 0.5 mile as it ascends a lightly forested step to Bluff Lake and a tiny companion pond. At the junction with Trail #90 just north of the lake, continue northwest on Trail #105 as it climbs along a splashing creek through a succession of small meadows sprinkled with wildflowers in midsummer. In 0.8 mile reach the junction of Trails #9 and #106 near a picnic area. Take a brief 1-mile side excursion north on loop Trail #106, the self-describing Flower Walk Trail, which rejoins Trail #9 slightly farther west.

Head west on Trail #9 as it weaves through ups and downs, gradually ascending to the upper end of Trail #100. Here turn south, and in a few feet cross a tributary of Crooked Creek. The song of splashing water may lure you off-trail a short distance downstream, where multiple braids of water cascade through a 20-foot-high wall of moss-encrusted boulders. In another 1,000 feet the trail descends to the base of Crooked Creek Falls, where the creek drops in a foaming ribbon over a 35-foot-high lava lip. The path then slips downhill through subalpine forest and meadows to return to the Bird Lake start point.

The second loop hike is shorter in distance but boundless in breathtaking overviews of the Hellroaring Creek drainage and the wild ridges and cliffs that frame it. Hike west from the Bird Creek Meadows parking lot on Trail #9 toward the picnic area. At a junction in 0.2 mile turn north on Trail #20, which soon bends west along the edge of an 800-foot-high wooded cliff above the south side of Hellroaring Creek. Every break in the trees opens new vistas of the deep glacier-cut valley, waterfalls cascading over lava cliffs, and the broad ice tongue of the Mazama Glacier above.

At 6,480 feet the path reaches a rocky viewpoint knob rimmed with scraggly mountain hemlock. The melting snout of the Mazama Glacier sends icy water twisting down black talus slopes to drop in ribbons over lava cliffs into the deep green valley below. A sharp-edged gravel ridge that traces a sinuous contour downslope from the glacier's end is the terminal moraine that marks the limit of the glacier's last advance. Stacks of short cliffs, separated by intervening talus slopes, are remnants of the layer-cake

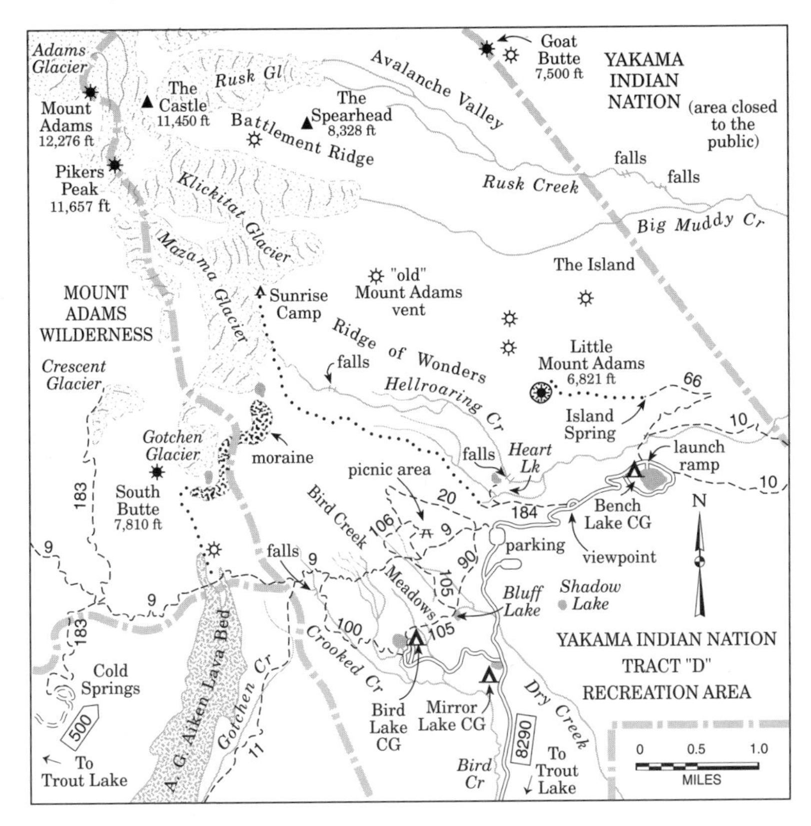

of lava flows that built the mountain. To the northwest, the vertical rock face of Battlement Ridge displays multiple bands of lava strata that run upward to the ragged ridge-top knob of the Castle.

On the north side of the valley, the eastward taper of the steep-walled rib of the Ridge of Wonders is interrupted by the blunt cinder cone of Little Mount Adams. East, the heavily forested Hellroaring Creek valley drops away to the canyon of the upper Klickitat River drainage, with views beyond to distant wheat fields of central Washington. When sated with the scenery, head south from the viewpoint on Trail #67, and in a short 0.3 mile reach the mid-point of the Flower Walk Trail at the Bird Creek Meadow Picnic Area. Trace either leg of this path back to Trail #9, then return east on it to the parking lot.

Crooked Creek Falls pours down a cliff of lava laid down when Mount Adams was formed.

Hellroaring Meadows

Hiking trail to a waterfall

Trail: Hellroaring Falls Trail #184
Rating: (E); hikers
Distance: Spring 0.8 mile, Heart Lake 0.8 mile
Elevation: Trailhead 5,260 feet, spring 5,360 feet, Heart Lake 5,373 feet
Maps: USGS Mount Adams East, USFS Mount Adams Ranger District
Driving Directions: Follow the driving directions above to the Bird Creek Meadows parking area in Tract "D." Continue east on the road to Bench Lake, and in 1 mile pull off into a side spur to the north to a viewpoint, helipad, and trailhead.

Take an easy, short walk into the scenic heart of the Hellroaring Meadows basin, encircled on three sides by 700- to 1,500-foot-high walls of a headwater cirque. At the trailhead viewpoint between the Bird Creek Meadows parking area and Bench Lake, start by savoring the view directly across the valley to the symmetrical pyramid of Little Mount Adams, one of the more impressive of Mount Adam's many parasitic cinder cones. The long avalanche-stripped wall of the Ridge of Wonders then leads the eye west to fields of glacial ice flowing from the false summit of Adams and high rugged cliffs that separate the ice falls on its east face.

Nearly flat Trail #184 heads west into this immense cirque, traversing a wooded slope for a little over ½ mile to a fork at the edge of Hellroaring Meadows. The left branch continues along the edge of the meadow for 0.2 mile to a spring feeding a tributary of Hellroaring Creek. The right branch crosses this creek and meanders through the meadow, with awe-inspiring views upward to ribbons of four snow-fed waterfalls dropping over 200-foot-high vertical cliffs in mid-slope. These cliffs and other similar bands above are outcrops of the various flows of overlapping lava that built the mountain over a period of more than 300,000 years. On the far edge of the meadow is a larger stream that, hopefully, can be crossed on a foot log (otherwise a frigid wade). From here a brief climb leads to the shallow bowl containing lovely little Heart Lake. Push a short distance east through the forest to glimpses through trees of the 100-foot-high cascade of Hellroaring Falls.

The view of Mount Adams from Hellroaring Meadows

Little Mount Adams

Hiking trail and cross-country route to a volcano crater

Trail: Island Spring Trail #66
Rating: (M), cross-country (D); hikers
Distance: 2.4 miles
Elevation: Trailhead 4,950 feet, summit 6,821 feet
Maps: USGS Mount Adams East, USFS Mount Adams Ranger District
Driving Directions: Follow the driving directions above to the Bird Creek Meadows parking area in Tract "D." Continue east on the road for 1.7 miles to Bench Lake Campground.

Aptly named, Little Mount Adams, on the shoulder of the Ridge of Wonders, is a miniature of its namesake, except for its lack of glaciers. This striking cinder cone, about 63,000 years old, is obviously of volcanic origin; it even has a small summit crater to prove it. Although there is no formal trail, the top can be reached from the east via a cross-country trek from an old sheep camp at Island Spring or from Hellroaring Meadows (preceding) via a longer and steeper cross-country hike that requires

The symmetrical cinder cone of Little Mount Adams is reflected in Heart Lake.

finding a way to cross Hellroaring Creek. The route via Island Spring is described here.

From the north side of Bench Lake Campground, Trail #66 switchbacks steeply down to a bridge crossing Hellroaring Creek. A hundred yards beyond the creek pass the junction with Trail #10, skirt the edges of a meadow, and start a moderately steep diagonal climb up the sidehill on the north side of the drainage. The tread is well maintained by a local elk herd working as trail crew for the Yakama tribe. At the nose of the ridge the trail swings west and wanders gradually uphill through open forest and pocket meadows. In about ½ mile arrive at the meadows at Indian Spring, the end of the formal trail. At this point the trail crew has run amuck, possibly from munching too much hellebore, and well-trod game trails go every which way, but nowhere in particular.

Cross-country travel isn't difficult, because the forest of true fir and mountain hemlock is fairly open. Just stay on the ridge top heading west, and eventually the timber thins to views of the target summit. The ridge top gradually changes from sparsely wooded meadows to steeper, loose, rust-colored tephra with the footing of a field of marbles. At the northwest rim of the crater, glance down into the 50-foot-deep cinder-covered bowl that was once an erupting volcano vent. A low wall of basalt encircles three-quarters of the crater, and on the northeast side a flat-topped outcrop marks the mountain's highest point.

The views more than repay the effort to get here. West, the narrow rib of the Ridge of Wonders drops away to deep, wide, glacier-cut cirques at the heads of Hellroaring and Big Muddy creeks—both streams born higher on the mountain's flanks from the Mazama and Klickitat glaciers. The jumbled black mass of the Klickitat Headwall divides two precipitous icefalls at the head of the Klickitat Glacier. On the north side of the glacier a vertical, 1,200-foot-high wall plummets from the narrow rock spine of Battlement Ridge. The ridge itself sweeps up unbroken from The Spearhead, 8,328 feet, to the ragged crown of The Castle, 11,450 feet, in a horizontal distance of just over a mile.

On the north rim of the Ridge of Wonders are a host of volcanic dikes that focus on the next ridge to the northwest, the location of the eroded core of the ancient cone of the predecessor to present-day Mount Adams. Two other knobs to the north on the Ridge of Wonders, and another northeast, mark volcano vents from 400 to 380 million years ago. Turn 180 degrees, and forested slopes drop and give way to grasslands, sagebrush, and wheat farms. Glenwood, 16 miles to the southeast, can easily be identified; just over the horizon is Goldendale, more then 45 miles away. To the south, in Oregon, the sharp white crown of Mount Hood probes the sky.

Take care on your return trip not to wander too far from the uphill path. A maze of game trails, some as well used as parts of Trail #66 itself, can muddy the process of finding the upper end of that trail at Island Spring. If not confident of your wilderness navigation skills, flag the route on the way up (and remove flags on the way down).

7

The Lewis River

The Lewis River originates high on the northwest side of Mount Adams at the base of the Adams Glacier. It flows in a westerly to southwesterly direction for more than 20 miles, takes a sharp bend to the south for about 7 miles, then resumes its southwesterly course along the south side of Mount St. Helens and eventually arrives at the mighty Columbia. Over time, the river has cut a canyon up to 800 feet deep and a mile wide through

Curley Creek Falls flows under a natural bridge of lava. Softer layers have been eroded away by the creek.

an ancient volcanic formation laid down between 26.8 and 19.5 million years ago. This formation is composed of many layers, each generally ranging from 3 to 10 feet thick, but occasionally forming thicker lens-shaped masses up to 600 feet thick. The composition of the individual layers is varied; some are made up of fine particles of volcanic debris welded together by heat, some are cobble-sized lava fragments imbedded in a softer mud flow matrix, others are sandstones formed from fine lava particles, and still others are eruptive lava flows, primarily of andesite.

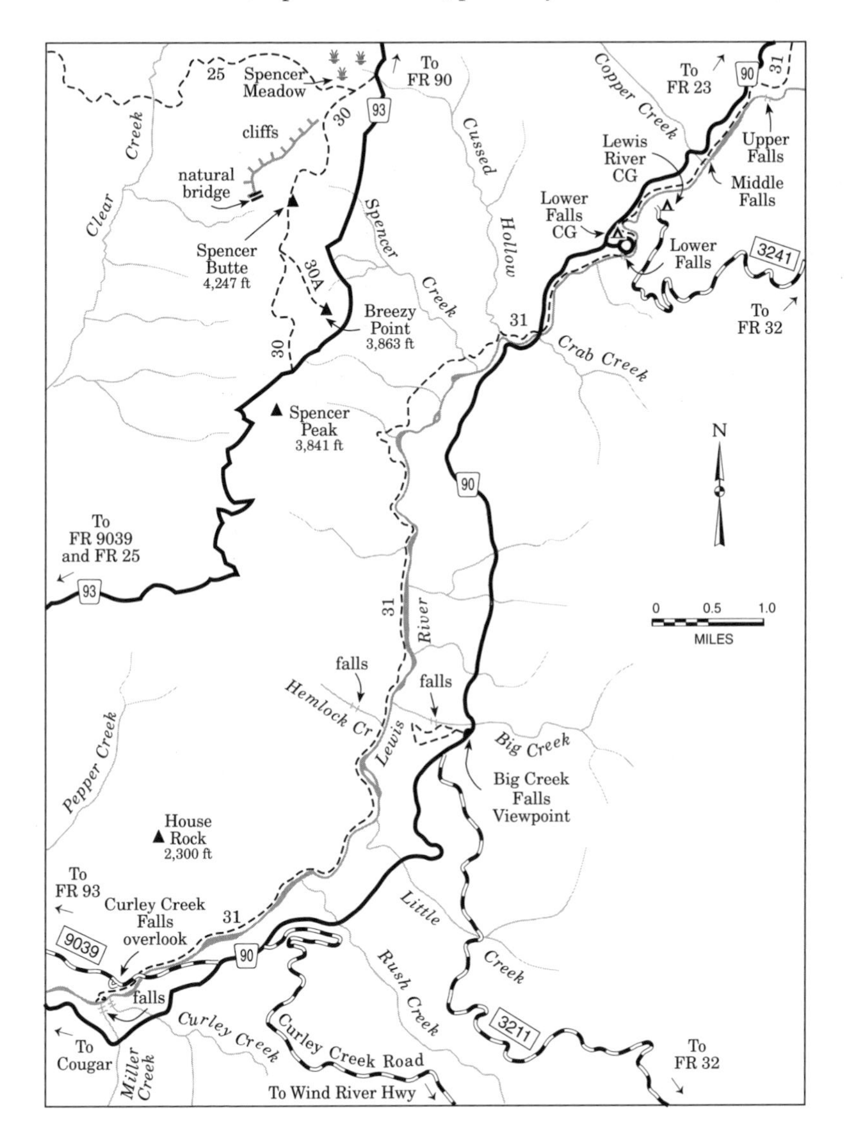

These various layers of rocks differ markedly in density and hardness. As a result, individual layers erode at different rates; the softer and more loosely structured rocks are washed away by the flow of creeks and rivers first, leaving behind the harder and more dense layers. This differential erosion, over time, creates dramatic features such as waterfalls and natural bridges. Some striking examples of these that are found in the Lewis River drainage are described here.

Curley and Miller Creek Falls

Short footpath and barrier-free path to waterfalls

Rating: Barrier-free (E); hikers
Elevation: 1,200 feet
Maps: USGS Burnt Peak, USFS Mount St. Helens Ranger District
Driving Directions: 5.2 miles east of the junction of FR 90 and FR 25 turn northwest on FR 9039 (1½-lane gravel), and in a mile take the turnoff to the left signed "Curley Creek Falls Overlook."

Here is a chance to view a pair of pretty waterfalls (in season), one combined with a natural bridge. Both are an easy stroll from a Forest Service viewpoint parking lot. All these features are a result of the differential erosion described in the chapter introduction. The Lewis River has worn away most of the large lava flow that once filled the canyon; only the edges of this flow remain in the cliff faces above its banks. Lateral streams such as Curley Creek have cut through softer strata in the remaining lava and have left standing the harder layers that form the bridge and the lips of the waterfalls.

The Curley Creek Falls Overlook is a mile off the major forest highway, FR 90, on the north side of the Lewis River. A level path, signed to the overlook, heads south into the brushy forest for 50 yards to an unsigned T-intersection. The trail heading upstream is the lower end of long Lewis River Trail #31. Take the branch heading downstream (Trail #31A); in another 150 yards a log fence rims an overlook high above the bank of the Lewis River. On the opposite side of the river, in the middle of the rock cliff, is a mouth-shaped cavity framed by a pair of moss-covered lips. The upper lip is a natural bridge approximately 6 feet thick and 20 feet wide; the lower one is about 60 feet above the river bank and generally has the white beard of Curley Creek spilling over it and splashing into the Lewis River. Late in a warm, dry summer the creek may dry up, leaving the mouth gaping.

The path switchbacks down to the bank of the river to yet another view of Curley Creek Falls, then continues 100 yards downstream to a viewpoint of Miller Creek Falls, also on the opposite side of the river. This narrow ribbon of water plunges over a rim of lava rock, then free-falls for 50 feet to the banks of the Lewis.

Spencer Butte Natural Arch

Hiking trail and cross-country route to a natural arch

Trail: Spencer Butte Trail #30
Rating: (M); hikers, mountain bicycles, saddle and pack stock, motorcycles
Distance: Either trailhead to the top of Spencer Butte 1.5 miles, top of the butte to the natural arch 0.3 mile
Elevation: South end of #30 3,430 feet, north end of #30 3,387 feet, top of Spencer Butte 4,247 feet, natural bridge 3,900 feet
Maps: USGS Spencer Butte, USFS Mount St. Helens Ranger District
Driving Directions: From the junction of FR 90 and FR 25, turn north on FR 25 (2-lane paved), and in another 6.2 miles head east on FR 93 (1-lane paved with turnouts). The road twists east, south, east, then north before arriving at the south end of Trail #30 in 7.9 miles. Continue north on FR 93 for another 3 miles to the north end of the trail.

This surrealistic arch of pressure-welded volcanic debris looks so precariously suspended that an unchecked sneeze would cause its collapse, yet it has probably stood here for hundreds of centuries and will most likely endure for many more. On a geological scale such permanence may be an unwise conjecture, however, as the arch, which fifteen years ago framed the nearby symmetrical cone of St. Helens, now frames only the ragged truncated stump of that peak.

Although the arch is fascinating to behold, no trail leads to it, and finding it is not a trivial task. Trail #30 starts and ends on FR 93; in between, the 3-mile trail crosses over the top of Spencer Butte. The top of the butte is 1.5 miles from either trailhead, and either approach involves about the same altitude gain. The trail from the north is recommended because it is the most scenic.

Cottongrass

The north end of the trail is located at the edge of cotton grass-covered Spencer Meadow, where you might spot elk, if you are lucky. The path starts as a jeep track, then soon becomes trail as it climbs through Douglas-fir and mountain hemlock forest with an understory of huckleberries. Unfortunately, the center of the tread has been deeply gouged by motorcycles in places, forcing hikers to place one foot directly in front of the other in the rut, in a gait that resembles a backwoods sobriety test.

In about a mile the path bends west to gain the open ridge-top meadow with continuous outstanding views east to Mount Adams, and west to St. Helens. In the heart of this meadow, at the high point of the butte (4,247 feet), are a group of concrete pads. These are the remnants of a fire lookout tower, built here in 1935, that replaced an earlier 1920s cupola cabin.

The search mission now begins; head due west toward Mount St. Helens, plowing through knee-deep huckleberry bushes and hopping over downed logs. In about ¼ mile the slope drops west, and young trees become more dense. Descend carefully because there is an abrupt cliff at about 3,800 feet; if you reach the top of it, you have gone too far. The arch is about 100 feet higher than the cliff and may require some side-slope traversing to locate. Finding your way back should be no problem—just head uphill due east. The trail runs north–south the length of the butte, so you'll eventually run into it.

Big Creek Falls and Hemlock Creek Falls

Short footpath and barrier-free path to views of waterfalls

Rating: Barrier-free to the Big Creek Falls viewpoint (E); hikers
Elevation: 1,800 feet
Maps: USGS Burnt Peak, USFS Mount St. Helens Ranger District
Driving Directions: From the junction of FR 90 and FR 25 continue east on FR 90 for 9.1 miles to the roadside pulloff at the Big Creek Falls Viewpoint parking area.

Tributaries of the Lewis River present two more enchanting waterfalls, one nearby and one across the river, for visitor admiration. An interpretive trail leads from the parking lot into a huge old-growth Douglas-fir and western red cedar forest, past massive root systems of several large trees exposed when these giants succumbed to winds and old age and crashed to the forest floor.

In 0.2 mile the path approaches the rim of a 200-foot-high vertical-walled canyon cut by Big Creek and first glimpses of the waterfall at its heart. Stay the temptation to sneak close to the canyon rim for closer peeks, for in a few hundred feet the path leads to an eagle-aerie observation platform with a clear view of this dramatic falls. Big Creek plunges in a foaming horsetail over a basalt lip for a 110-foot drop into a deep pool at its base. Behind the falls the cliff retreats in the shape of an inverted bowl with two major bands of volcanic strata clearly exhibited in its wall.

The trail continues west for another 0.5 mile through more magnificent old-growth to arrive at a rock outcrop atop a 400-foot-high cliff that overlooks the sinuous path where the Lewis River has cut through these ancient volcanic strata. Across the river to the west, Hemlock Creek can be seen as it slips over old basalt flows and plummets in a free-falling horsetail for over 200 feet to lower cascades that descend to the Lewis River.

Lower Lewis River Falls

Short footpaths and barrier-free path to falls overlooks

Rating: (E), barrier-free; hikers
Elevation: 1,520 feet
Maps: USGS Spencer Butte, USFS Mount St. Helens Ranger District
Driving Directions: From the junction of FR 25 and FR 90, continue east on FR 90 for 14.1 miles to FR 9054; follow this short spur, which goes east into the Lower Falls Campground and Recreation Area.

Lower Lewis River Falls has been called a "miniature Niagara Falls"—not because of its height or volume of water, but because of its form and general appearance. The falls spread in a wide fan over a massive, 100-foot-wide block of basalt that spans the river from bank to bank. The water drops about 35 feet over a series of narrow shelves that were formed within the larger block by overlapping lava flows. The result is not a single, monolithic waterfall, but a wall of water with a dozen or more separate cascades, ranging in size from slender trickles to the roaring flood of the primary river course.

Above the falls, the Lower Falls Recreation Area fills a forested flat at the heart of a horseshoe bend in the river. In addition to viewpoints of the falls, the recreation area has picnic sites and a campground that is perhaps one of the best and most pleasant in the Gifford Pinchot National Forest.

Lower Lewis River Falls cascade over a series of rock shelves.

8

Dark Divide

The name Dark Divide most precisely describes the ridge crest north of Dark Peak. However, it often is more loosely applied to the Dark Divide Roadless Area, a larger region southwest of Randle between the Cispus River (on the north and east), the Lewis River (on the south), and Iron Creek (on the west). Although logging roads invade its drainages, and clearcuts scar far too many of its slopes, the high country in the heart of this area presently is the largest roadless section of the Gifford Pinchot National Forest outside of its dedicated wilderness areas. The three major components of the Dark Divide are the long, high ridge running north from Dark Mountain through Jumbo Peak, Sunrise Peak, Juniper Peak, and Tongue Mountain; a parallel crest to the west from Holdaway Butte through McCoy Peak to Langille Peak; and the strip of forested ridges that straddle the east–west Boundary Trail #1 between Table Mountain, on the east, and Elk Pass, on the west.

McCoy Peak (far right) intrudes strata of Langille Ridge. The rock in the foreground is on the edge of an old volcano vent.

The geology of this area is varied and complex. The oldest geological formation found here is between 26 and 24 million years old. It is a thick composite of layer upon layer of extruded lava, mud flows, lava fragments of various sizes and shapes imbedded in a loose matrix, and sandstones formed from volcanic material. The upper portions of this formation make up most of the walls of the McCoy, Yellowjacket, Badger, and Pinto creek drainages up to an elevation of about 3,600 feet.

Overlaying this formation is another one composed of stacks of lava flows, welded eruptive debris, airfall volcanic deposits, and sandstones of volcanic origin that accumulated over a 7-million-year period, ending some 19 million years ago. This formation makes up most of Langille Ridge above 3,600 feet and the north end of the ridge between Juniper Peak and Tongue Mountain. Most of the remaining country above 3,600 feet on the south and west sides of the Dark Divide is comprised of a sandwich of andesite lava flows, bedded layers of andesite lava fragments, and a few basalt lava flows dating 25 to 18 million years ago.

A number of prominent rock outcrops are found atop the ridges and in valley walls in portions of the McCoy and Yellowjacket creeks that are not part of the preceding formations. Instead, these are intrusions of magma into fissures and weak spots in these formations, such as a major fault line running along the bottom of the McCoy Creek drainage. These intrusive lavas are typically diorite or dacite, a much harder rock than that found in the formations they penetrate. As a result, when the softer layers are carried away by erosion, these harder cores remain behind, forming

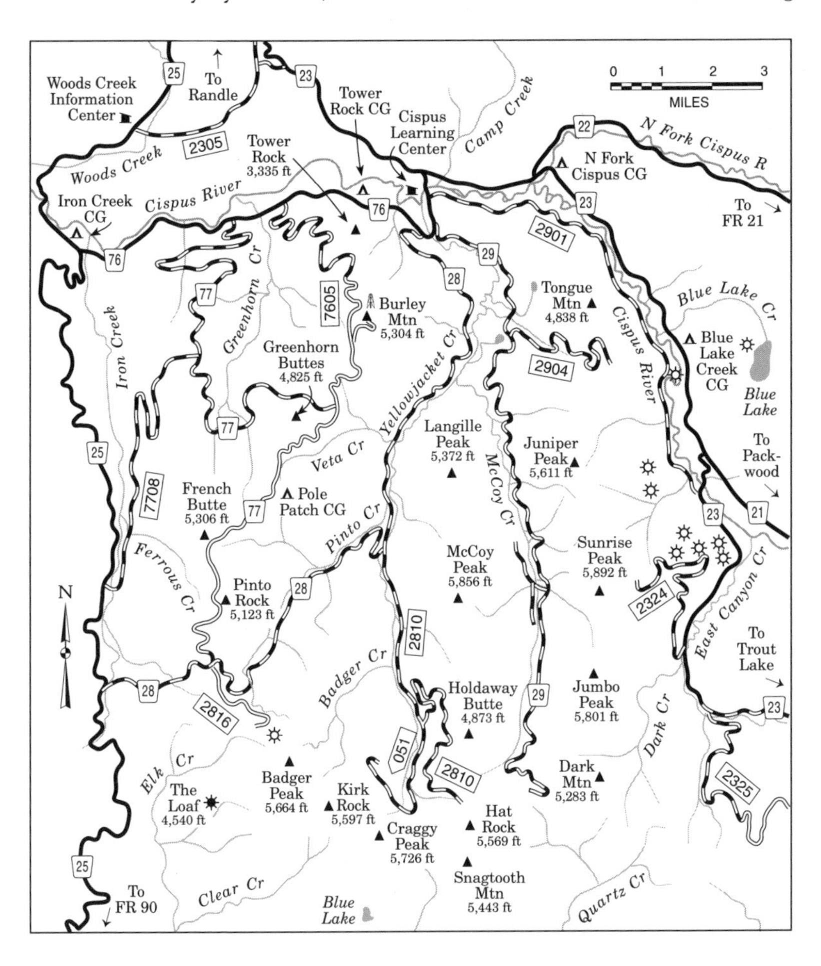

prominent rock exposures such as Sunrise Peak, Tongue Mountain, McCoy Peak, Tower Rock, and Shark Rock.

The only relatively new formation in the area is the long tongue of basalt that forms the top of broad Badger Ridge between Pinto and Yellowjacket creeks. This lava was extruded to the northeast about 650,000 years ago from a vent at point 5195, atop the ridge immediately south of the end of FR 2816 (the beginning of Trail #257); it then filled the valley below, where it ponded and cooled. Pinto and Yellowjacket creeks later cut deep canyons along the sides of the flow, thereby inverting the topography, leaving a basalt plateau in place of the original valley. Major earthquakes have also modified the terrain, causing minor cliff-creating landslides in several spots and an extremely large slide of more than 2 square miles in size that peeled from the saddle south of Tongue Mountain, then spread across and filled side-slope canyons to the northwest all the way down to Yellowjacket Creek.

Badger Peak

Hiking trail to a viewpoint of lava flows, debris, and a dacite intrusion

Trails: Badger Ridge Trail #257, Badger Peak Trail #257A
Rating: (M); hikers, mountain bicycles, saddle and pack stock, motorcycles
Distance: 1.3 miles
Elevation: Trailhead 4,880 feet, Badger Ridge 5,300 feet, #257A 5,220 feet, summit 5,664 feet
Maps: USGS French Butte, USFS Randle Ranger District
Driving Directions: From US 12 at Randle, head south on Highway 131 and in 1 mile bear south on FR 25 (2-lane paved). Continue south for another 21.8 miles, then turn east on FR 28 (1½-lane gravel), signed to Mosquito Meadows. In 2.8 miles, at the junction with FR 77 and FR 2816, head south on FR 2816 (1-lane dirt) for 4.6 miles to the road-end trailhead.

This old fire lookout site commands a view of over 700 square miles of the beautiful rugged peaks, ridges, and valleys in the heart of the Gifford Pinchot National Forest, as well as a horizon lined with the giant glacier-clad volcanoes of the South Cascades. Alas, the lookout built in 1924 is no longer there—it was destroyed by the Forest Service in 1960.

There are various approaches to Badger Peak; the shortest one is from the north via Trail #257. The road to the trailhead is a challenge for a passenger car. It is rough and narrow and, although annually graded, the tread becomes deeply eroded by run-off and traffic from motorcycles and four-wheel-drives.

Even the trailhead has striking views, although they are limited to features to the north: Rainier, Langille Ridge, McCoy Peak, the Goat Rocks, and the surrealistic outcrop of Pinto Rock. Point 5195, the knob immediately south of the trailhead, is a vent for a major basalt flow forming the high interdrainage plateau to the north.

The trail begins in sparse mountain hemlock with hellebore, lupine, and huckleberry bushes lining the way. It soon breaks out beneath a 200-foot-high vertical cliff where it threads through massive chunks of rockfall from above. Although the rockfall is not recent, it still serves as motivation not to tarry. The tread has a loose coating of tephra from the 1980 eruption of St. Helens that makes the steep climb to the top of Badger Ridge slick and tedious, but ever-expanding views over McCoy Peak to Juniper Ridge compensate the effort.

The shallow green pool of Badger Lake lies on the west side of Badger Peak.

The ridge-top respite is brief, for now the track drops down the other side at an equally steep grade. At 0.8 mile is the intersection with Trail #257A, which—what else—begins another steep struggling sidehill climb with the same loose footing. Near the summit the path crosses a steep chute that is often filled with snow until late summer and can be dangerous to cross when icy. The path finally reaches the bare, exposed summit knob in 0.5 mile, and here the views! The ridge to the southeast pauses first at Kirk Rock, whose bulbous summit outcrop resembles a gear shift knob overhanging cliffs below. Farther east the slab-sided fin of Shark

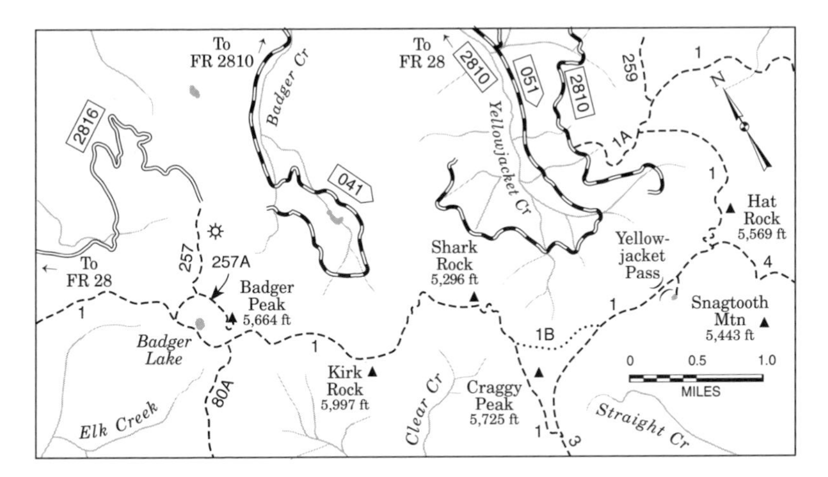

Rock knifes through ridge-top trees, and south from it the wooded crest rises to rock teeth atop Craggy Peak.

Northeast, across the deep Yellowjacket Creek drainage, note the strata that dip north gradually in the side slopes between McCoy and Langille peaks. These are exposures of the volcanic flows and debris that built the ridge, layer by layer, over a 7-million-year period. On the ridge beyond, the long cliff-faced crest of Juniper Ridge is capped by the summits of Jumbo Peak, made up of andesite lava flows, and the intrusive dacite thumb of Sunrise Peak. The bare high country of the Goat Rocks lines the horizon. On each quadrant are the premier cones of the Cascades: Rainier to the north, St. Helens to the west, Hood to the south, and Adams to the east.

Shark Rock Scenic Area

Hiking trail past diorite intrusions

Trails: Badger Ridge Trail #257, Boundary Trail #1
Rating: (M); hikers, mountain bicycles, saddle and pack stock, motorcycles
Distance: To Craggy Peak 4.3 miles
Elevation: Trailhead 4,880 feet, at Badger Ridge 5,300 feet, junction of #257 and #1 4,930 feet, at Kirk Rock 5,000 feet, at Shark Rock 5,260 feet, at Craggy Peak 5,020 feet
Maps: USGS French Butte, McCoy Peak; USFS Randle Ranger District
Driving Directions: Follow directions to Badger Peak, preceding.

Boundary Trail #1 was laid out in the early 1900s to mark the dividing line between what was, at that time, the Rainier and Columbia National forests. Aging gracefully, as befits a well-kept senior citizen among trails, it makes its way through moderately open old-growth forest, turning here to touch a lovely wooded lake, bending there to pass a particularly scenic view, always maintaining a moderate grade as it twists across side slopes and along ridge tops in its long east–west journey. Only a generous fall-out of debris from the 1980 St. Helens eruption disrupts the otherwise solid tread with a 2- to 6-inch-thick layer of loose, slippery, pea-sized tephra.

For a reasonable day-trip along the Boundary Trail into the Shark Rock Scenic Area, take Badger Ridge Trail #257, described earlier, and continue down another 0.5 mile past Trail #257A to the junction with the Boundary Trail near the shore of the shallow green pool of Badger Lake. It would be an absolutely beatific spot, were it not for a nearby sidehill ripped up by macho motorcycle jockeys pursuing god-knows-what gratification.

Follow the trail east past tempting glimpses through forest of the overhanging tip of Kirk Rock, then suddenly burst into an open meadow right below its striking north face. The wall that rises over 200 feet above is clearly separated into three wide diagonal bands. The upper and lower ones are of a harder andesite, while the intermediate layer is composed of a softer conglomerate of volcanic debris. The entire face appears to have been formed by the north side of the peak splitting away in a huge landslide.

Streams of smoke from a distant forest fire, borne by 60 m.p.h. winds, drift eerily past peaks of the Shark Rock Scenic Area. Shark Rock is the fin-like summit on the left, Craggy Peak is center left, and Kirk Rock is the nearby knob.

The more rapid erosion of the softer band of rock across its midsection has left the upper layer overhanging. Pick out a hillside cave at the lower east side of the middle band, and speculate (or explore, if gutsy) what might be residing in it.

The path ducks back into thick forest and in another 0.5 mile crosses a rib, where a few hundred yards north of the trail a group of short rock towers thrust out about 80 feet above the crest. This rock differs from that exposed in the face of Kirk Rock in that the strata are more vertical than horizontal. This is understandable, because the outcrop is the erosion-exposed tip of a vertical intrusion of diorite into the base rock in the area. The steep slabby side of another such intrusion that forms Shark Rock lies 0.5 mile farther along the ridge to the southeast. The trail crosses the center of this formation, then switchbacks down its 200-foot-high east face.

Here catch views of the snaggle-toothed top of Craggy Peak, at 5,725 feet the highest of the mini-peaks in the Scenic Area. The path proceeds along the crest to the base of Craggy Peak, but the forest cover blocks closer views. For the day-hiker, this is a good turnaround point. The Boundary Trail is now routed around the west and southeast sides of Craggy Peak, but at one time there was also a segment, no longer maintained, that crossed beneath the north face of the peak. If adventuresome, a good backcountry navigator might try to find and follow it to complete a 3-mile loop around the base of the peak. Hikers that make prior arrangements for drop-offs and pick-ups can continue east to Boundary Trail accesses from the heads of Yellowjacket and McCoy creeks.

Pinto Ridge

Road trip to a volcanic outcrop

Elevation: Base of rock 4,660 feet, summit 5,123 feet
Maps: USGS French Butte, USFS Randle Ranger District
Driving Directions: Follow driving directions to Badger Peak, earlier, as far as the junction with FR 77, then turn north on FR 77 (1-lane dirt). In 2.2 miles reach a pulloff at the south base of Pinto Rock. FR 77 can be followed north for 21.1 miles to paved FR 76, on the south bank of the Cispus River.

The buttresses and battlements of the ridge-end fortress of Pinto Rock make it one of the most distinctive features in the Dark Divide. The top is over 1,000 vertical feet above drainages to the south and west and 1,800 feet above the one on its east side. The south side of the massive rock outcropping rises over 250 feet above its immediate base, and its upper east face overhangs the wall below. A deep cleft separates the bulk of the castle from a smaller outwork to the north.

Forest Road 77 offers a close-up view of massive Pinto Rock.

Their placement and elevation above the surrounding landscape make both Pinto Rock and Pinto Ridge readily identifiable from any high spot within a 4- to 5-mile radius to the east, south, or west. For a close-up and even more impressive view, turn north from FR 28 onto FR 77, and follow this narrow, rutted dirt road steeply uphill for nearly 2 miles. On rounding a corner, the rock suddenly comes into full view immediately in front of you. In another 500 yards find a pulloff at the base of the rock and, if you wish, explore the perimeter via a cross-country scramble. Several climbing routes lead to the top of the rock, but none are easy; all are rated Class 5 or higher. Do not attempt any of them unless you have advanced climbing skills and are properly equipped with rock-climbing gear.

Even a casual observer will notice several horizontal layers in the face of the peak. Some of these strata are fine-grained sandstones and siltstones formed from lava particles by pressure; others are beds of larger volcanic debris imbedded and compressed in a matrix of similar origin. The vertical towers and grooves that lace the faces of the rock are caused by wind and water erosion that have attacked its soft surface since the rock was formed 18 to 16 million years ago. The volcanic fragments that compose the rock and the ridge to its north are thought to have originated from a complex of vents somewhere between Pinto Rock and Greenhorn Buttes, 3 miles to its northeast. Several andesite dikes exposed near the rock and on the ridge north were intruded in the same time frame and are likely from the magma source that fed the vent complex.

Pinto Ridge extends north for over a mile from the ridge-end rock, with several smaller outcrops exposed by erosion along its crest. FR 77 continues north along the west side of the crest; however, this is not a road for the faint-of-heart. It is narrow, replete with ruts and potholes, with no protection on the outside edge where it drops away abruptly for an eternity (actually 1,000 feet) to the valley floor below. Once past the north end of the ridge, the road improves slightly and for the next mile offers knock-out views of ridges and valleys to the east.

Tower Rock

Road trip past an intrusive sill

Elevation: Base of rock 1,400 feet, summit 3,335 feet
Maps: USGS Tower Rock, USFS Randle Ranger District
Driving Directions: Head south from US 12 at Randle on Highway 131, and in 1 mile bear south on FR 25 (2-lane paved). Turn east in 8.9 miles onto FR 76 (1-lane paved with turnouts). Pass the base of Tower Rock in 6.3 miles, and reach Tower Rock Campground in another 0.5 mile.

Impressive to look at, but a bear to get up—that's Tower Rock! FR 76 runs east–west between the Cispus River and the rock, about ¼ mile away from its base. From the west, the road threads through a dense cover of second-growth forest; the rock first comes into view as the road crosses open pastures of farms just below its face. A short way east of the entrance to Tower Rock Campground another spur on the north side of the road leads to openings with a profile view of the rock face as it rises 1,100 vertical feet above its forest footings.

Those driven by demons to attempt the north face should be aware that it is an extremely difficult climb, rated as Class 5.8, requiring advanced climbing skills, ropes, and artificial aids for assistance. Most should be satisfied with marveling at it from the road below.

A glance at this 12-million-year-old rock shows none of the often-seen layer cake of horizontal lava flows. This is because the rock is the flattened end of a sill—magma that was forced between the bedding planes in the overlying layers of rock and solidified before breaking through the surface. Later erosion of the softer surface layers enclosing it exposed the hard diorite tower we see today.

Phantom Falls and Angel Falls

Barrier-free forest path and hiking trails to waterfalls

Trails: Covel Creek Trail #228, Phantom Falls Trail #228A, Angel Falls Loop Trail #228B, Burley Mountain Trail #256

Rating: Barrier-free loop (E), #228, #228A, and #256 (M), #228B (D); hikers

Distance: Covel Creek 1.2 miles, barrier-free 0.4 mile, Phantom Falls 1 mile, Angle Falls Loop Trail 3 miles

Elevation: Trailhead 1,280 feet, junction with #228A 1,390 feet, Covel Creek Falls 1,780 feet, #228B high point 2,250 feet

Maps: USGS Tower Rock, USFS Randle Ranger District

Driving Directions: From US 12 at Randle, head south on Highway 131 for 1 mile, then bear east on FR 23 (2-lane paved). In 8.4 miles head south on FR 28 (2-lane paved), then turn west in 1.5 miles onto FR 76. Reach the Cispus Learning Center in 0.7 mile. The Forest Service prefers that hikers park just outside the entrance gate on the west side of the entrance triangle, near the welcome sign.

Three different trail loops are linked to provide a spectrum of hiking experiences for visitors at the USFS Cispus Learning Center southeast of Randle. The trails permit exploration of various physical, geological, and biological features of the Covel Creek drainage south of the center.

The lowest and easiest loop is 0.4 mile long and flat; a guide rope strung between posts along one side assists vision-impaired visitors. The path starts and ends at the pedestrian crosswalk 300 yards east of the entrance to the center and passes various forest features such as root systems of fallen trees. A bench along the way invites you to pause briefly and listen to the whispers of water trickling in an adjacent creek.

The second loop trail, Trail #228, which leaves FR 76 opposite the Learning Center entrance, is rated Easy. It begins as an old jeep road, wide enough

The trail, which runs behind Phantom Falls, gives hikers an "inside view" of the stunning waterfall.

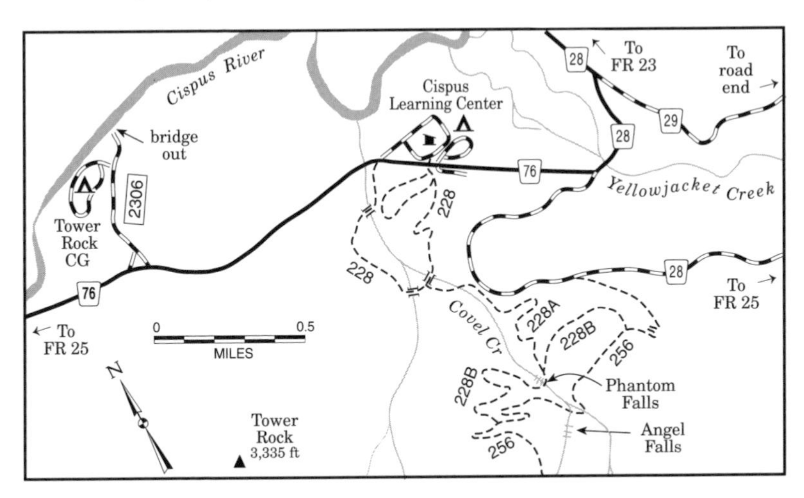

to walk two abreast. The path weaves through open Douglas-fir forest carpeted with Oregon oxalis, crosses a creek bed (dry in late summer), then follows its west bank gradually uphill for 0.8 mile to a junction with Phantom Falls Trail #228A. Trail #228 continues east across another bridge to an adjacent lean-to picnic site, then switchbacks gently downhill and in 0.5 mile reaches the main road at the pedestrian crosswalk.

After leaving Trail #228, Phantom Falls Trail #228A climbs briefly to yet another bridge, this time over flowing water, then heads into the steep walls of the rapidly narrowing drainage. A very stiff switchback pulls away from the gully, then a second returns the path to a rib about 150 feet above the creek. Both creek and trail climb steadily; the trail takes more switchbacks to climb above a precipitous drainage wall. Several sparkling cascades can be seen in the stream below.

Angel Falls Loop Trail #228B, the third (and more difficult) loop trail, is met at a T-junction in 0.5 mile. Just west of the junction the trail bends into a fern-, moss-, and lichen-clad pocket behind striking Phantom Falls. The 60-foot-high falls drops over the lip of an andesite lava flow in a veil so wide and thin that the view from behind is much like looking through a car windshield in a rainstorm.

The fork east of the junction diagonals up a steep face, then circles the upper rim of a wooded cliff, with intermittent views out across the Cispus River valley. It meets Burley Mountain Trail #256 in another 0.5 mile. The west fork, after emerging from behind Phantom Falls, strikes uphill with a vengeance. Its narrow, rough tread climbs across a very steep, exposed slope; after two switchbacks and a 0.5-mile ascent, it too arrives at Trail #256. To complete the loop, follow Trail #256 downhill for a mile to its junction with the lower fork. En route recross Covel Creek, with views uphill to the ribbon of Angel Falls cascading down the rock face of the narrow canyon.

McCoy Peak

Hiking trail to the top of a dacite intrusion

Trails: Rough Trail #283, Langille Ridge Trail #259, McCoy Trail #259A
Rating: (D); hikers, mountain bicycles, saddle and pack stock, motorcycles
Distance: 4.3 miles
Elevation: Trailhead 2,680 feet, summit 5,856 feet
Maps: USGS McCoy Peak, USFS Randle Ranger District
Driving Directions: From US 12 at Randle, head south on Highway 131 for 1 mile, then bear east on FR 23 (2-lane paved). In 8.4 miles head south on FR 28 (2-lane paved), then in 1.1 mile bear southeast on FR 29 (1½-lane gravel). Turn west on FR 2900115 (1-lane gravel) in 10 miles to reach the start of Trail #283 in 0.2 mile.

McCoy Peak is the highest point on Langille Ridge. It commands an outstanding view east across the deep canyon cut by McCoy Creek to the rugged crest of Juniper Ridge as it runs through Jumbo, Sunrise, and Juniper peaks, Tongue Mountain, and several more unnamed pinnacles. To the west, Pinto Rock, the rock outcrops atop Pinto Ridge, the bald protrusion of Burley Mountain, and the steep-sided wedge of the basalt plateau between Yellowjacket and Pinto creeks all cry out, "This is vertical country!" At the headwaters of Yellowjacket Creek, dark, heavily wooded ridges are broken by the cliff faces of Badger and Craggy peaks and Kirk, Shark, and Hat rocks. The familiar volcanic giants of Adams, St. Helens, and Rainier, as well as the Goat Rocks, line the distant skyline.

The old fire lookout that stood atop the peak between the 1930s and 1960s is gone, thus the path from the ridge trail to the summit is no longer maintained and may be hard to follow. However, the rewards are worth the effort, as you feast on the panorama to the point of visual overload. The shortest, albeit not the easiest, route to the summit is from McCoy Creek via Rough Trail #283. This name is not in jest, because the track takes endless, extremely steep switchbacks up a densely wooded knife-edge ridge, gaining over 2,700 feet of elevation in 1.7 miles—that's going uphill folks!

At the ridge-top junction with Trail #259, stop and rest. Then, after a short ridge-top romp, lose a third of that hard-gained elevation as the path switchbacks abruptly down the headwall of Bear Creek to skirt the lower end of a buttress that runs east from the top of McCoy Peak. Another climb northwest gains a 5,220-foot-high saddle on the northwest flank of the peak where the summit path takes off. This vague trail diagonals south up the forested slope to the ridge top, then follows it east as it hooks to the summit of McCoy Peak.

At view breaks along the long trail climb, look north along the side slopes above the McCoy Creek drainage, where avalanche tracks and bare

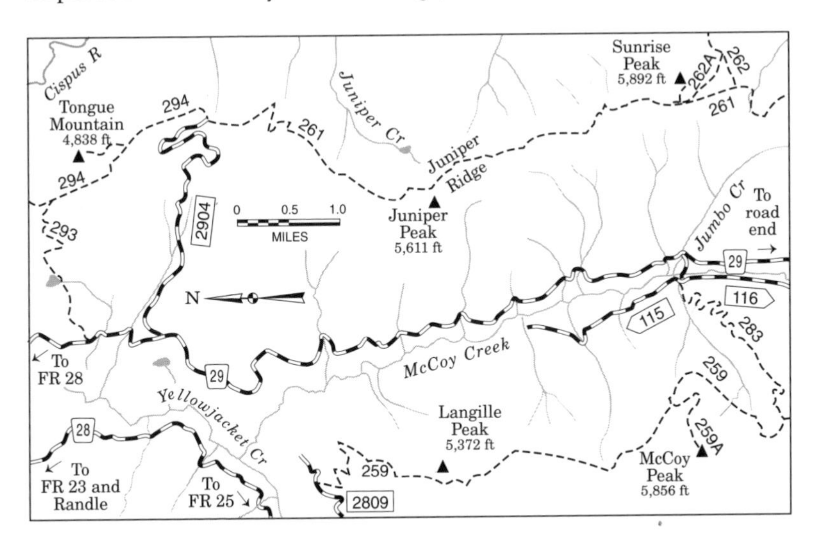

patches expose the faint lines of layer upon layer of lava flows and compacted volcanic debris; all taper gradually downhill to the north. This accumulation of volcanic material stacked up to build the ridge over a 7-million-year period, starting about 26 million years ago. However, at the rocky summit of McCoy Peak hard blocks of dacite are tilted upward to near-vertical, indicating that the peak was formed from a magma intrusion into the overlying strata. The softer surface rock was eroded by millions of years of cutting action by glaciers and creeks on the flanks of the peak. A host of dikes surrounding McCoy Peak appear to have been feeders for a large ancient volcano that was entirely eroded away during the Cascades uplift.

Tongue Mountain

Hiking trail to the top of a diorite intrusion

Trails: Tongue Mountain Trail #294, Tongue Mountain Summit Trail #294A
Rating: (M); hikers, #294 mountain bicycles, motorcycles
Distance: 1.7 miles
Elevation: Trailhead 3,620 feet, summit 4,838 feet
Maps: USGS Tower Rock, USFS Randle Ranger District
Driving Directions: Follow directions to McCoy Peak, preceding, as far as the turn onto FR 29. From FR 29 turn east on FR 2904 (1-lane gravel) in 3.9 miles, and reach the trailhead in another 4.2 miles

Tongue Mountain sits in the jaw of Juniper Ridge like a giant rock molar. From the Cispus River the east face of the peak rises nearly 3,300 vertical feet in a distance of less than a mile. Forest hides the angularity of the lower 2,000 feet, but the remainder of its east face sheds this disguise to reveal a sheer rock wall rising to a double summit.

The altitude gain is nearly the same from the McCoy Creek drainage on the west side of the mountain, but the topography is less dramatic. Sometime within the last half-million years a deep-seated earthquake triggered a major landslide that started near the saddle north of the peak and swept down across its base, filling in its lower slopes and creating a broad bench that drastically modified the apparent steepness of the peak. Its former contours are seen only in the rock cliffs in the upper 800 feet of the peak. These are quartz diorite remnants of a 25-million-year-old magma intrusion that have been exposed by erosion.

Trail #294 leaves FR 2904 at the saddle north of the peak, then heads north in a leisurely ascent through moderately open forest of fir and hemlock. The path hits the ridge top and levels off for a bit, then both ridge and trail resume the uphill grade, the ridge at a much more rapid rate. The sideslope ascent remains in timber, frustrating anticipated views. The junction with summit spur Trail #294A is reached in about 1 mile, where log barricades enforce the "hikers only" sign on the uphill trail.

The route climbs steeply to rejoin the ridge top, then finally breaks into the long-anticipated treeless slope below the summit cliffs. A narrow tread snakes upward in a chain of switchbacks to the cleft between the blocks that crown the mountain. This was the end of the trail for pack stock that supplied a former lookout site atop the peak. The lookout was staffed periodically between 1934 and 1947. The reason the trail stops will be obvious if you peer over the cliff rim a few feet to the east. A stomach-churning two-thirds of a mile straight below is the blue ribbon of the Cispus River. The route to the summit, ahead, is less a trail than a strenuous scramble up a steep, rocky face, where a misstep could be disastrous.

The erosion-resistant rock of Tongue Mountain is part of an ancient magma intrusion.

Even from the trail-end col the views are outstanding—the only feature missed by passing up the summit is Mount Rainier. Across the valley to the west, a forested ridge tapers north from the bald pyramid of Burley Mountain down to the blunt turned-up nose of Tower Rock. Southwest is rough-topped Langille Ridge; beyond is yet another long ridge that ends in cliff-walled Pinto Rock; in the background is Mount St. Helens. The green crest of Juniper Ridge stretches south, capped by the vertical-faced fingers of Juniper and Sunrise peaks. East, the wall of the lower summit frames the flank of Mount Adams.

Juniper Ridge

Hiking trails to views of various volcanic features

Trail: Juniper Peak Trail #261
Rating: (M); hikers, mountain bicycles, saddle and pack stock, motorcycles
Distance: 3.4 miles
Elevation: Trailhead 3,620 feet, Juniper Ridge 5,420 feet
Maps: USGS McCoy Peak, Tower Rock; USFS Randle Ranger District
Driving Directions: Follow directions to Tongue Mountain, preceding.

Technically, Juniper Ridge is only the ½-mile portion of the long, high north–south crest between McCoy Creek and the Cispus River that is immediately south of Juniper Peak. However, general usage tacks the name on the entire 8-mile-long ridgeline between Tongue and Dark mountains. Aside from lower portions in the adjoining drainages, most of this ridge has thus far eluded chain saws; pray that it will continue to do so.

The first stretch of trail is a pleasant forest hike. View country is reached about 2 miles south of the FR 2904 trailhead, when the trail gains the sparsely wooded crest high on the shoulder of Juniper Peak. A brief dodge east of the ridge top to avoid bands of cliffs leads to a rock-ribbed notch immediately north of the vertical-to-overhanging tip of Juniper Peak. After crossing through this col, the path traverses the steep bare east side of the peak to the narrow rock rib of the "real" Juniper Ridge running southeast from the summit to a pyramid-shaped point ½ mile away. Cliffs east fall away from the crest to glacier-carved cirques at the head of Juniper Creek. West, grassy slopes drop to precipitous, timbered walls at the head of Camp Creek.

This open section of ridge is a grand viewing platform for the splendid scenery of the north part of Dark Divide. On its west side is the V-shaped canyon cut by McCoy Creek through 3,000 vertical feet of accumulated lava flows, beds of consolidated pumice and volcanic debris, and sedimentary rock formed from fine particles of eroded lava. Bare spots in the sidehill slopes expose individual pages banded into chapters in this 20-million-year-old book of the region's volcanic history. Below and upstream from the ridge Camp and Scamp creeks, two unnamed companions between them, and Jumbo Creek all drain along a series of lateral local faults linked with a major fault line running down the bottom of the valley. A dacite intrusion into the same area is the result of magma flow upward through these cracks in the underlying crust. It may be the plug remaining from a massive volcano vent that is thought to have erupted here 20 to 15 million years ago. No trace of the volcano remains, other than the intrusive dacite.

Across the valley a rugged, equally high ridge is capped on the south by McCoy Peak and on the north by the large, jumbled outcrop of Langille Peak. South from Juniper Ridge, along the crest and its flanking buttresses, are two prominent summits, Sunrise and Jumbo peaks, and an assortment of cliff-framed pinnacles. Sunrise is a distinctive fang of intruded dacite with a 200-foot-high, vertical wall on its east face and a steeply tapered open slope below to the west. A fire lookout camp that first perched on its summit in 1924 was replaced by a more conventional tower in 1934; the tower was destroyed by the Forest Service in the 1960s. Jumbo is a massive pillbox-shaped summit with an 80- to 200-foot-high lava flow band wrapped around three sides. North, Rainier rises above the shoulder of Tongue Mountain. The glacier-traced bulk of Adams dominates the southeast horizon, and the jagged top of the Goat Rocks Wilderness serrates the northeast skyline.

The Cispus River Drainage

The main flow of the Cispus River originates in the heart of the Goat Rocks Wilderness; major tributaries from the south start as frigid melt-water from glaciers on Mount Adams. The north fork of the river begins in high ridges bounding the west side of the main fork drainage. The 50-square-mile, heart-shaped area between the two forks of the river is composed of a series of contorted ridges rising above wide, glacier-cut valleys. The core of this area rises abruptly from a river bank level of 1,700 feet to ridge-top elevations ranging between 4,900 and 5,600 feet. Lower fringes of this zone have seen heavy logging, and chain saws are rapidly gnawing upward through beautiful old-growth forest still mantling upper slopes.

The bulk of the region is composed of old andesite flows interbedded with layers of ash, tephra, and volcanic fragments in a matrix of similar origin. These multiple strata were deposited over a period from 38 to 25 million years ago; subsequent glaciation has bared this layer-cake of rock bands in slopes beside many streams. Aside from a number of basalt and dacite intrusions along the perimeter of the region near the end of this period, the most unique geological feature is a relatively young (300,000 to 150,000 years old) basalt volcano that erupted on Blue Lake Ridge, damming the stream to the north to create the area's watery centerpiece, Blue Lake.

Flows from an ancient volcano dammed a stream to create Blue Lake.

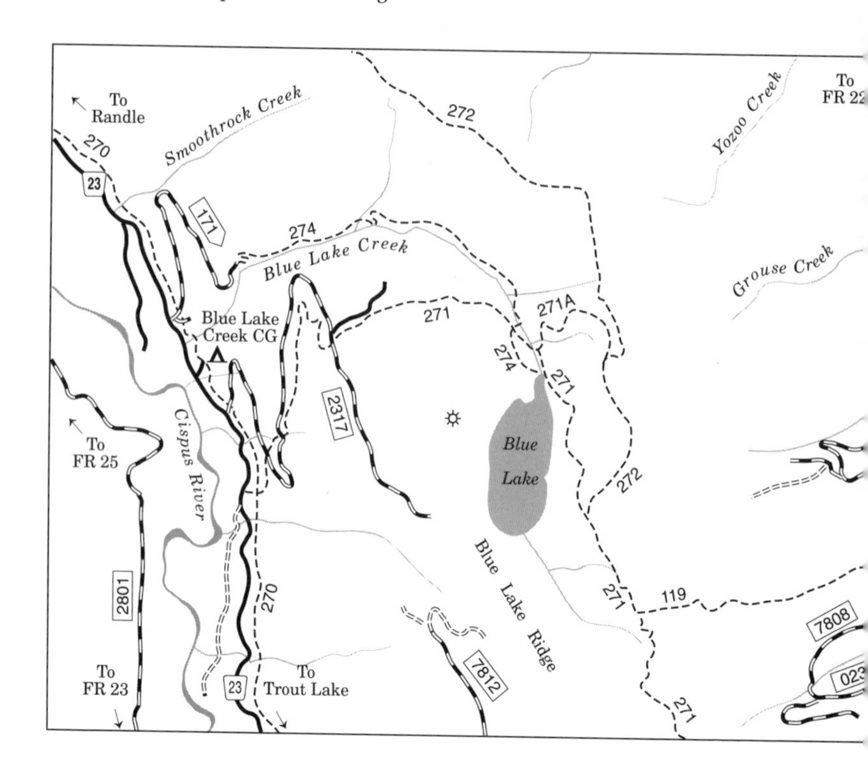

Blue Lake

Hiking trail past a lava flow

Trail: Blue Lake Hiker Trail #274
Rating: (D, X); hikers
Distance: 2.5 miles
Elevation: Trailhead 2,400 feet, Blue Lake 4,058 feet
Maps: USGS Blue Lake, USFS Randle Ranger District
Driving Directions: From US 12 at Randle, head south on Highway 131 for 1 mile, then bear east on FR 23 (2-lane paved). In 15.3 miles turn east on FR 2300171 (1-lane gravel) and in 3.3 miles arrive at the road-end trailhead.

A trail built in 1993 leads past one of the most spectacular walls of columnar basalt in the South Cascades. However, you must pay the price to get there, because the route is narrow and painfully steep, with points of unnerving exposure. The road to the trailhead is no piece of cake either; it too is narrow and exposed as it crosses the top of a clearcut just below the road end. In its favor, the clear side slope provides knockout views across the Cispus River to Juniper Ridge and the cliff-faced summits of Tongue Mountain and Juniper and Sunrise peaks.

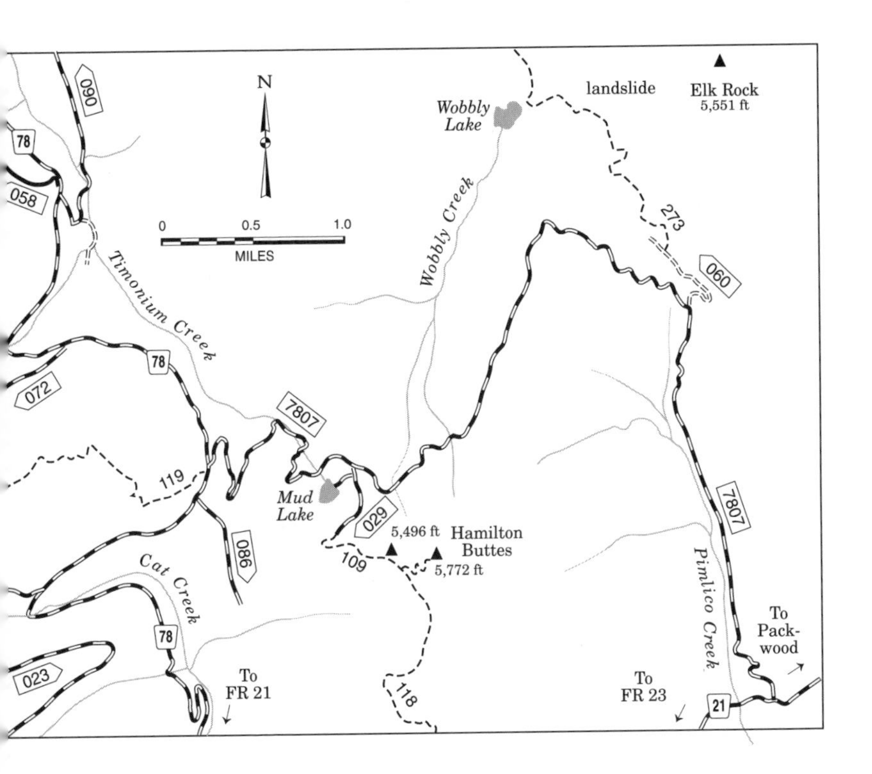

The trail, which begins in dense forest, high on the north flank of the Blue Lake Creek drainage, knows only one direction—up! It is so steep in places that rock staircases have been built around protruding outcrops; just 30 inches from these rock slabs the slope drops abruptly for over 200 feet. The path crests briefly after a gain of 250 feet, then, as fast as it climbed, it descends to the creek bed 50 feet below, at the base of an overwhelming, 250-foot-high wall of columnar basalt.

A lower skirt of perfectly parallel columns, each about 8 inches in diameter, runs along the base of the wall. At one place a dark cave has been eroded into its base. About 50 feet above this orderly vertical array the wall transforms into swirling plates and columns, going in all directions, with moss and lichens streaking the light gray basalt with pale greens and yellows. The huge wall is a surrealistic mosaic created from a pool of magma by an eruptive artistic genius.

This basalt was extruded from a volcano vent above the west side of Blue Lake. The flow created the lake when it blocked the canyon below. But why is there a 250-foot-high wall of lava on one side of the creek, and no trace of anything comparable on the other side 30 feet away? The current theory is that at the time the volcano erupted, glaciers still covered the area. The basalt initially erupted beneath the glacial ice, creating a hole that was filled with melt-water, which quickly quenched the hot lava. The

This wall of columnar basalt can be seen on the way to Blue Lake.

result was the steep-sided pedestal of rapidly solidified basalt. The volcano eventually broke through the ice, continued to grow with more basalt extrusions, and the vent was finally capped by a cinder cone.

Columns such as these in the wall normally form in basalt when it shrinks and splits as a result of rapid cooling. This would be the case if the flow pooled up against a wall of ice. When the climate warmed, the glaciers disappeared and the creek flowing through the canyon eroded away the softer layers of rock to the north to expose the wall we see today.

How long ago did this occur? The best guess is that it was during the second, most recent period of glaciation. Unfortunately, this doesn't resolve the volcano's age, because geologists have differing opinions on when that glacial period occurred; estimates range from 60,000 to 300,000 years ago.

The massive lava wall is certainly sufficient reward for the strenuous effort to reach it, and one could not be faulted for turning back at this point. However, the steepest and most exposed parts of the trail are behind. It resumes its upward trek, first through dense cedar and hemlock, then in more open forest. More rock staircases snake around outcrops as the path keeps pace with the rising creek bed, staying between 50 and 100 feet above its narrow, steep-walled channel.

At about 3,900 feet, both trail and creek level, and in 0.2 mile the path crosses a bridge and intersects the wide motorcycle speedway of Trail #271, just a few hundred feet beyond the latest logging operation scalping the southwest side of Blue Lake Ridge. The route continues on through light timber for another 0.2 mile to the shores of Blue Lake. The sandy shore tapers into the turquoise depths of the lake. Thoughts of a skinny dip disappear with first touch of the icy water.

Hamilton Buttes

Hiking trail to views of volcanic dikes and landslides

Trail: Hamilton Peak Trail #109
Rating: (M); hikers, mountain bicycles, saddle and pack stock, motorcycles
Distance: 1 mile
Elevation: Trailhead 5,120 feet, summit 5,772 feet
Maps: USGS Hamilton Buttes, USFS Randle Ranger District
Driving Directions: From US 12 at Randle, head south on Highway 131 for 1 mile, then bear east on FR 23 (2-lane paved). In 18.1 miles, at the junction of FR 23 and FR 21, continue southeast on FR 21 (1-lane paved with turnouts). In another 6.9 miles head north on FR 78 (1½-lane gravel for 1.5 miles, then 1-lane gravel), follow it as it twists uphill for 5.3 miles, then turn east on FR 7807 (1-lane gravel) and in 1.9 miles south on spur FR 78007029 (1-lane gravel) signed to Mud Lake. The road is blocked just beyond a parking area in another 0.6 mile.

A short hike leads to an old lookout site that combines panoramic views usually associated with such locations with equally interesting looks at several striking geological features. The lookout cabin that was built on the summit in 1924 has suffered the fate of nearly all of its companions; it was abandoned in the 1960s, and no longer exists.

The first introduction to the geology of the region comes on the road into the area. As FR 78 bends around the Cat Creek drainage, massive clearcuts expose a 30-foot-wide band of rock that runs horizontally the length of the north wall of the drainage, about 500 feet above the roadbed. This is one example of the multiple layers of ancient lava flows, lava flow debris, and volcanic sandstones that accumulated here between 36 and 30 million years ago. They underlay most of the area between the Cispus River and its north fork. Although logging is currently responsible for removing the forest and making this strata visible, it was probably first exposed by local alpine glaciers that cut deep into the base rock to shape most of its major drainages. It was further shaped by subsequent erosion by creeks that run through these channels.

As FR 78 reaches the saddle between Cat and Timonium creeks, the treeless upper slopes of the long ridge to the northeast are conspicuously striped with band upon band of narrow rock outcrops dipping gradually downslope to the north. This is yet another example of the layering of volcanic deposits underlying the area.

From the saddle, continue on FR 7807 and spur 029 to the bulldozed parking area, as described in the driving directions. Although the road at one time continued another 0.5 mile uphill, a large mound of dirt prevents driving beyond the parking area. The steep section above is designated as Trail #109.

As the trail climbs the side slope east toward the summit, openings in the forest offer views north to the Goat Rocks; these are meager samples compared to what you will see from the top. In 0.5 mile the road/trail meets the head of Trail #118 from the south, then tapers to a path as it continues up to the bald summit ridge. Rainier, Adams, and Hood are all presented in full glacial uniform, while the rough backbone of the nearby Goat Rocks Wilderness is clad in dull brown combat dress, adorned with swatches of snowfields only in early summer. On the north side of the deep cleft cut by the North Fork of the Cispus, green forest blotched by logged-off patches rises to the rocky ridge top of Twin Sisters, Castle, Cispus, and Cold Springs buttes and Smith and South points.

Trails on the rock slopes of Hamilton Mountain are ideal for mountain biking.

The sandwich of rocky bands noted earlier on the west slope of the ridge north from Hamilton Buttes emerges again on its east slope above Wobbly Creek. However, an elongated rock fin midway along the ridge, and the vertical west face of Hamilton Mountain itself, are anomalies that run almost perpendicular to that ridge. They are not a part of the basic underlying formation but are volcanic dikes of andesite that intruded through weaknesses in its structure some 7 million years after it was formed. The dike at Hamilton Mountain also extruded a thin flow of andesite that covers an area of about 200 acres on the slope southeast of the summit.

The upward fold that created the ridge north from Hamilton Mountain continues beyond the North Fork of the Cispus as Stonewall Ridge. From the summit of Hamilton you can see a dramatic repetition of the 6-million-year accumulation of layers of volcanic debris in the sheer, 700-foot-high wall of Stonewall Ridge. Landslides contributed to the geological history of the surrounding landscape, as well as glaciers and volcanic extrusions. The ridge northeast from Hamilton Buttes is capped by Elk Peak; a close look at its west face can pick out traces of a large landslide that swept down into Wobbly Creek, damming it to form Wobbly Lake.

10

Mount Adams North

Glaciers on the north and northeast slopes of Cascade peaks such as Mount Adams become larger and survive longer than those on south- and west-facing slopes because they are subjected to less direct sunlight. The larger the glacier, the more intense its erosive power, thus the upper slopes on the north side of Mount Adams are much more rugged and precipitous than those on the south flank. Glacial till and moraine deposits, extending far downslope from the tongues of present-day ice flows, mark past advances of the mountain's glaciers. Near the foot of these upper slopes

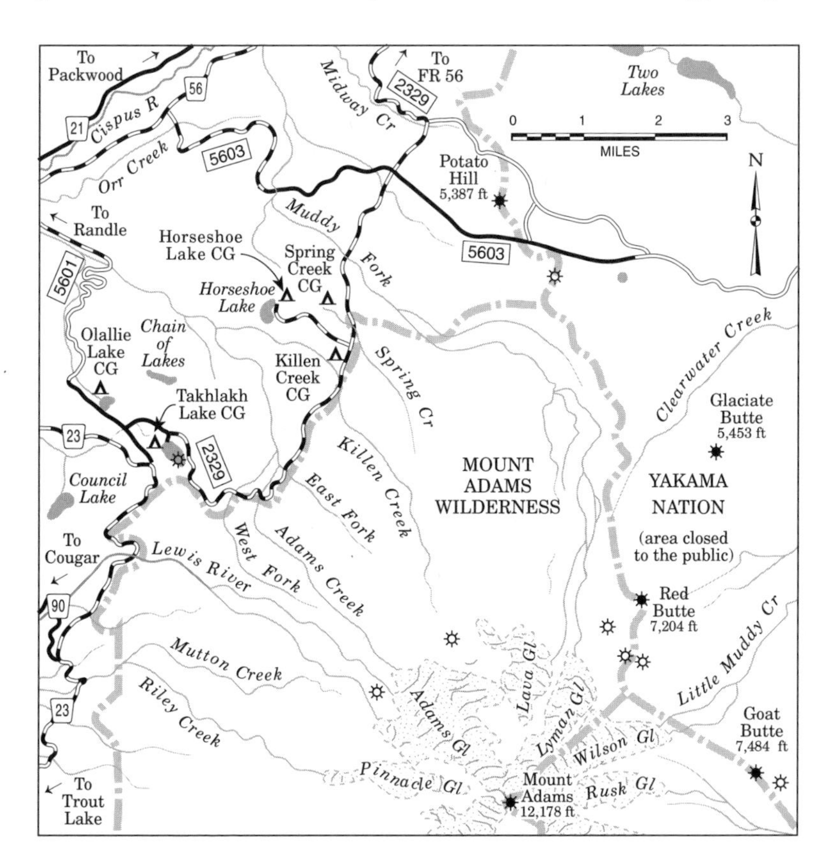

Two distinctive volcanoes: the cinder cone of Potato Hill in the foreground and the glaciated stratovolcano of Mount Adams in the distance

subalpine forest covers a broad, rather uniform mile-high plateau formed from younger lava flows. These flows originated from vents higher on the mountain, then spread gradually downhill, like tongues of honey, over older rock formations. Several younger eruptions, many within the last 12,000 to 15,000 years, have built distinctive cinder cones or large fields of blocky andesite and basalt that have not yet been obscured by impinging forest growth.

Takhlakh Lake

Hiking trail around a lake of volcanic origin

Trail: Takhlakh Loop Trail #134
Rating: (E); hikers
Distance: 1.1 miles
Elevation: Takhlakh Lake 4,385 feet
Maps: USGS Green Mountain, USFS Randle Ranger District
Driving Directions: The lake can be reached in three ways. From Randle, leave US 12 at Randle and head south on Highway 131 for 1 mile, then bear east on FR 23 (2-lane paved). In 18.1 miles, at the junction of FR 23 and FR 21, continue southeast on FR 21 (1-lane paved with turnouts) for 4.8 miles, then turn southeast on FR 56 (1-lane paved). In 0.5 mile head south on FR 5601 (rough 1-lane gravel) for 5.7 miles, where it becomes 1-lane paved at the Olallie Lake access road. In another 0.5 mile turn east on FR 2329 (1-lane paved) to reach Takhlakh Lake Campground in 1 mile.

From Packwood, leave US 12 2.5 miles southwest of Packwood, turn south on FR 21 (2-lane gravel), and in 16.6 miles head east on FR 2160 (1-lane paved) for 1.8 miles. Here take FR 56 southwest for 1.2 miles, then bear south on FR 2329 (both 1½-lane gravel). Arrive at Takhlakh Lake Campground in another 14.1 miles

From Trout Lake, head north from Trout Lake on Highway 17, signed to Glenwood, and at a junction in 2 miles continue north on FR 23 (1-lane paved with turnouts), which becomes 1½-lane gravel in 7.4 miles. Continue north on FR 23 for another 15.6 miles to FR 2329, turn east on it, and reach Takhlakh Lake Campground in 1 mile.

The azure surface of Takhlakh Lake forms a nearly perfect, circular mirror filled with the reflection of the massive northwest face of Adams and the fractured, icy tongue of the Adams Glacier dropping precipitously from the summit cone. A forest of subalpine fir rims the lake. A large forest camp and boat launch ramp are found along its west shore. Brazen forest inhabitants show no fear of campers: chipmunks scamper across picnic tables at the first sign of food, trying to beat the competition of gray and Steller's jays to any handouts—voluntary or not. A small flotilla of eared grebes may be seen paddling the lake with studied indifference toward boats and shoreside fishermen. A pleasant shoreside trail makes a 1-mile loop around the lake; it also connects to a spur at its south rim that leads down to Takh Takh Meadow Trail (see following).

Takhlakh and its nearby companions, Olallie Lake and Chain of Lakes, sit on a small 4,400-foot-high timbered plateau in pockets carved by glaciers into the surface of one of the few dacite flows on the perimeter of

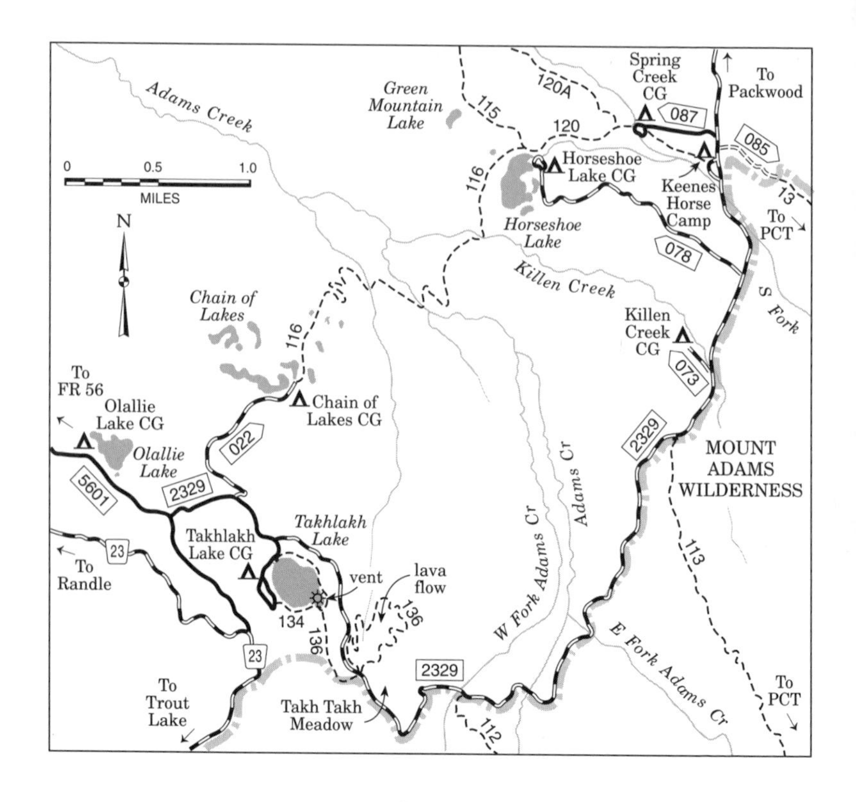

Mount Adams. This lava is believed to have been extruded during one of the earliest phases of volcanism that started to build Mount Adams, probably 385,000 to 375,000 years ago. Recent geological mapping indicates that the southeast end of the depression in which Takhlakh Lake lies may be the source vent for the flow.

Takh Takh Meadow

Hiking trail through a lava flow

Trail: Takh Takh Meadow Trail #136
Rating: (E); hikers
Distance: 1.5 miles
Elevation: Trailhead 4,550 feet
Maps: USGS Green Mountain, USFS Randle Ranger District
Driving Directions: Follow directions to Takhlakh Lake Campground, preceding, and take FR 2329 (1-lane gravel) southeast for 1.1 miles to the trailhead.

An easy loop hike offers stark contrasts in terrain, ranging from a moonscape canyon of blocky lava boulders to a flat, flower-clad meadow. Both

A field of blocky lava edges the trail in Takh Takh Meadow.

are bordered by a forest of tall, slender subalpine fir. The loop trail can be reached from a wooded extension that joins Takhlakh Loop Trail #134 near the lake's south shore or from a roadside trailhead where this spur crosses FR 2329 at the northwest end of Takh Takh Meadow.

The road parallels the long 400-foot-wide meadow for nearly ½ mile southeast from the trailhead, separated from it by a thin fringe of trees. No formal trails lead through the meadow itself. If you leave the trail to wander through its knee-deep grass to admire and photograph the array of wildflowers that bloom here from midsummer to early fall, use care not to trample the posies (and be prepared to fend off mosquitoes).

Traveling counterclockwise, the loop trail leaves the northwest corner of the meadow, then traces the base of a 50-foot-high field of andesite blocks about 50 to 100 feet east of FR 2329. In about 300 yards the path twists across the fractured lava bank, then wends northward atop the edge of the flow, which now drops into a steep, north-flowing drainage. Confusing side paths lead to choice viewpoints along the route. The tread is fill dirt that has been dumped into the cracks and chinks in the jumble of lava blocks to make a relatively smooth surface; the path on either side is a maze of fractures, holes, and channels, some as much as 10 feet deep, between surfaces of huge boulders that comprise the flow.

The track diagonals down, across the lumpy flow into the center of the drainage, then works back up through the black boulders to the rim on the opposite side. After dipping into another shallow channel, the way snakes uphill and drops off the lava bed to the edge of Takh Takh Meadow, which it follows west to the start point of the loop.

The lava flow that you have just toured is relatively new, probably only 10,000 to 4,000 years old—just yesterday in a geological time scale. The flow is one of several extruded from vents low on the flanks of Adams within the last 10,000 years. Although no single distinctive source for this lava bed has been identified, it probably originated from a vent between 700 and 1,000 feet higher up the slopes to the southeast.

Muddy Fork Lava Flow

Hiking trail past a lava flow and volcano vents

Trails: PCT #2000, Muddy Meadows Trail #13, High Line Trail #114
Rating: #2000 (E), #13 (M), #114 (D); hikers, saddle and pack stock
Distance: Toe of the flow 1.2 miles, Devils Gardens 11 miles
Elevation: Trailhead 4,750 feet, toe of the flow 4,600 feet, Devils Gardens 7,760 feet
Maps: USGS Glaciate Butte, Green Mountain, Mount Adams East; USFS Randle and Mount Adams ranger districts
Driving Directions: Leave US 12 2.5 miles southwest of Packwood, turn south on FR 21 (2-lane gravel), and in 16.6 miles head east on FR 2160 (1-lane paved) for 1.8 miles. Here take FR 56 (1½-lane gravel) southwest for 1.2 miles, then bear south on FR 2329 (1½-lane gravel). In 5.6 miles turn southeast onto FR 5603 (2-lane paved), and in 1.7 miles reach the parking area where Trail #2000 crosses the road.

Sometime between 6,800 and 3,500 years ago a massive flow of blocky andesite was ejected from a vent on the northwest slope of Mount Adams. The vent was at the 7,200-foot level of the mountain, about ½ mile southwest of Red Butte. The lava swept north, then west, for more than 7 miles down the Muddy Fork drainage. An easy day-hike south along the PCT provides the opportunity to explore the toe of this flow. A much longer and more difficult trip that probably requires a backcountry overnight camp can take you high on the northeast shoulder of Adams to barren slopes near the flow's source, just off Trail #114, north of Devils Gardens. The latter trip also reaches the terminal moraines of the Lyman and Lava glaciers and opens dramatic views of the seldom-visited east side of Mount Adams.

For the short day-trip, take the PCT south from where it crosses FR 5603 at the parking area. An easy downhill grade heads through a dense forest of mountain hemlock and subalpine fir and in 1.2 miles reaches the edge of a 50-foot-high bank of blocky lava, sparsely covered with brush and trees. This is the lower end of the lava flow described earlier. Follow the trail for another 0.8 mile as it skirts the nose of the blocky lava field or, if in an adventuresome mood, climb on the flow itself and probe the fractured surface to see the plants that have established themselves in this rough environment over the past 2,000 years.

Beyond this point the route swings south away from the lava bed, and doesn't reapproach it for another 8 miles and 1,600 feet of elevation gain. This is a good turnaround spot, unless you plan to make this at least a two-day trip. If you choose the long backpack, continue south on the PCT, cross a pair of branches of the Muddy Fork, then start a ceaseless uphill

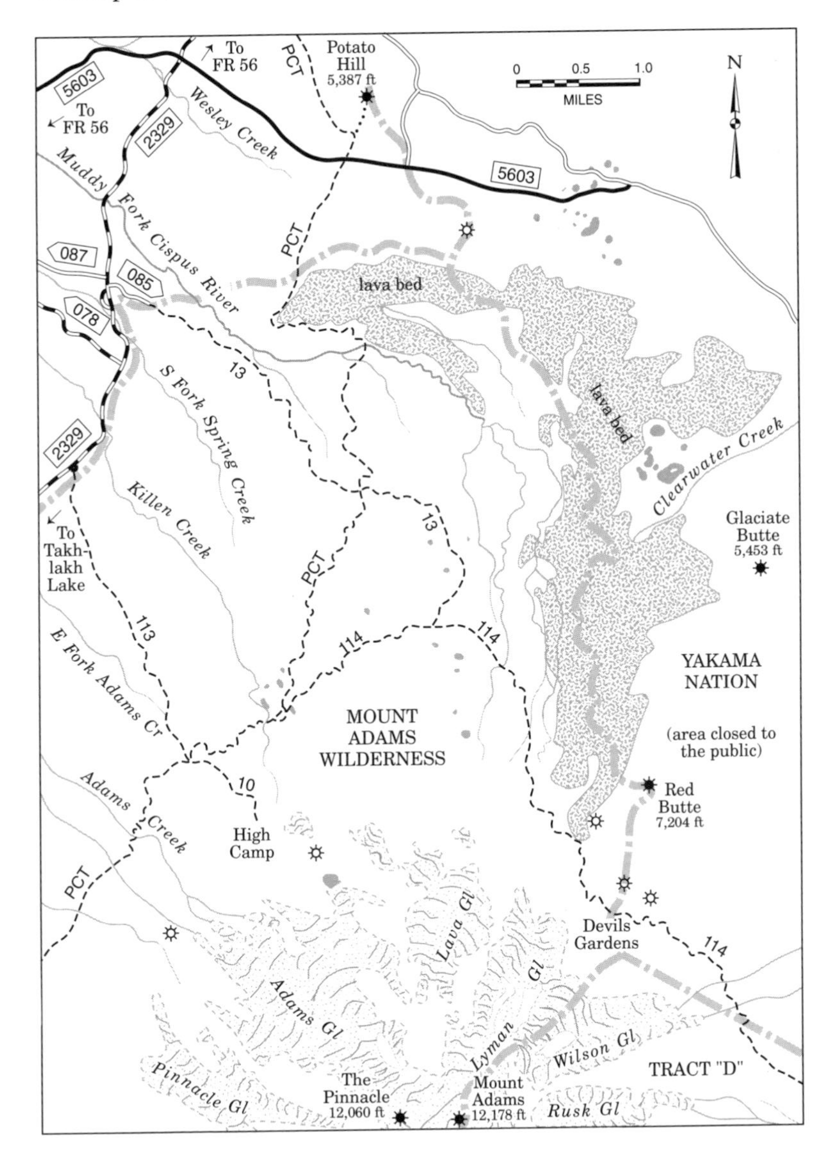

slog through thick subalpine fir and lodgepole pine for nearly 2 miles to a junction with Trail #13. Head east on Trail #13, which soon bends south as it traces a shallow stream bed up through moderately open forest to an intersection with Trail #114 at 5,860 feet. Head east on #114 for another mile to Foggy Flat, 6,000 feet. Here, next to a creek flowing into the Muddy Fork, is probably the last decent campsite protected by trees and with ample water.

In another 0.5 mile reach timberline. Here are sweeping views up the Lava Glacier to the rock and ice bowl of the Lava Headwall, at the convergence of North Cleaver and Lava Ridge. The path is now across either snow or barren rock and glacial till, depending on the weather and time of year. As the route wends upward, it crosses meltwater streams from the Lava and Lyman glaciers that feed the Muddy Fork; these may prove a challenge during periods of heavy run-off. At 7,200 feet the cone of Red Butte is northeast, with its shallow summit crater exposed through a gap in its badly eroded northwest lip. The shallow depression about 100 feet below and 200 yards northeast of the trail is the source vent for the Muddy Fork lava flow.

The trail continues across the no-man's-land below the terminal moraine of the Lyman Glacier to the broad, wind-swept boulder patch of Devils Garden, a mile north of Red Butte. Above, icefalls of the Lyman and Wilson glaciers plunging from the summit snowfields of Adams are framed by sawtooth ribs of lava rock. The crest of the Goat Rocks Wilderness fills the skyline to the northeast. This plateau saddle is an excellent turnaround spot because the rugged trail beyond drops steeply to the east—altitude that would have to be regained on the route out.

Potato Hill

Hiking trail and cross-country route to a volcano cone

Trail: PCT #2000
Rating: #2000 (E), cross-country (M); hikers, saddle and pack stock
Distance: #2000 0.4 mile, cross-country 0.5 mile
Elevation: Trailhead 4,750 feet, Potato Hill summit 5,387 feet
Maps: USGS Glaciate Butte, Green Mountain; USFS Randle Ranger District
Driving Directions: Follow directions to Muddy Fork Lava Flow, preceding.

No question at all about the volcanic origin of Potato Hill. Its cone shape stands out strikingly atop a relatively flat subalpine plateau. Its isolation also means that it is an ideal spot from which to scan the surrounding countryside. From the roadside trailhead at FR 5603, the PCT goes northeast, headed for the heart of the Potato Hill cinder cone looming above the trail a scant 0.8 mile away.

In 0.2 mile the PCT bends sharply northwest to skirt around the base of the hill. Here, start a cross-country ascent of the remaining 600 vertical

Hikers will discover a shallow crater at the top of the cinder cone of Potato Hill.

feet to the summit. Routefinding is not difficult, because there are only sparse low trees and brush on the south slope, and the summit rim is nearly always in sight. The main concern is to avoid trespassing into the Yakama Indian Reservation; the unmarked boundary bisects the summit from southeast to northwest. The slope is too steep to comfortably climb head-on, and the footing is a loose composition of orange-, yellow-, and rust-colored clinkers and tephra, so a linked series of short switchbacks is the best technique to reach the top. Use established game trails whenever possible.

On reaching the rim, you will discover a small, shallow, meadow-filled bowl that is the one-time crater in the heart of the volcano cone. Here, about 110,000 years ago, thin floods of basalt lava were disgorged across the valley to the north, filling it to a depth of about 90 feet. The north rim of the crater is covered by subalpine fir, limiting views to the north; however, the south edge has only a few scrub trees to interfere with the easy ascent to the high spot of the hill on the east side of the rim.

Views south and east are wide open. Nothing but a low forest carpet stands between here and the immense mass of Mount Adams. The lava plain spreads to the southeast, alternately covered by green forest or barren logged strips, then it drops into the deep chasm cut by the Klickitat River over tens of thousands of years. To the northeast, the sharp tips of a portion of the Goat Rocks are seen above an intervening ridge at the headwaters of the Klickitat River.

11

The Goat Rocks

The remote, rugged terrain of the Goat Rocks Wilderness is marked by sharp summits, cliffy walls, rocky spires, and deeply incised valleys. Several small glaciers lie on peaks at its heart. The wilderness, which straddles the Cascade crest south of White Pass, is a 105,023-acre subalpine-to-alpine parkland, with elevations running from 2,200 to 8,184 feet.

The geological history of the Goat Rocks area is as fascinating as its rumpled scenery. From high points in the area and on its perimeter, one can look back over 140 million years of evolution of the Cascade Range. Visualize a giant composite volcano, similar to, and only slightly smaller than, present-day Mount Rainier, that developed in the heart of the Goat Rocks over a period from 3.2 to 1 million years ago. Looking at today's eroded remains of this once-magnificent volcanic peak, the mind's eye can imagine what mighty Rainier will probably look like some 1 to 2 million years in the future.

The oldest rocks in the area, which date from 140 to 120 million years ago, are presently exposed in lower walls of the deep drainages cut by the Tieton River. These were formed off the coast of North America from accumulated volcanic sediments and major basalt lava flows beneath the surface of the sea and were carried to the region by movement of the oceanic plate. Volcanic eruptions over millions of years built layers of up to 10,000 feet thick. Beginning about 17 million years ago, these older rocks were uplifted, and they folded and fractured in roughly a north–south direction. Erosion of more recent overlying rock has exposed this older formation in the northwest end of the wilderness in the Upper Lake Creek basin and in the headwaters of the Clear Fork of the Cowlitz River and the North Fork of the Tieton River.

Between 3.8 and 2.8 million years ago fresh magma intruded through fractures in the older rocks, forming several thick north–south-oriented dacite dikes in the northeast corner of the wilderness. During this period, explosive volcanic eruptions near Devils Horns spread flows of rhyolite and volcanic debris fragments into the headwaters of the North and South Forks of the Tieton River. A second vent in the same spot, at Devils Washbasin, later erupted localized basalt lava flows. The remnants of this eruption today make up the pinnacles of Devils Horns.

The volcanic activity that built the Goat Rocks Volcano began about 3.8 million years ago with andesite extrusions from a complex of vents stretching between what are presently Gilbert and Johnson peaks. More

Gilbert Peak, shown here, and other spires of the Goat Rocks are the remains of an ancient stratovolcano. Mount Adams is in the distance.

than fifty andesite dikes around the edges of this area are directed toward the pipes that fed the growth of the volcano. Enough time elapsed that deep drainages were cut in the initial lava flows, so later eruptions coursed down deep river-cut valleys, filling them with masses of andesite. Subsequent erosion cut new channels around the edges of the valley-filling flows, leaving them as today's high plateaus such as Pinegrass Ridge and Bear Creek Mountain.

With the eruption of the Hogback Mountain volcano, between 2 million and 300,000 years ago, volcanic activity began anew. A vent just north of Shoe Lake erupted olivine basalt lavas over a 3-mile area that extended north to present-day White Pass, and a lava shield more than 2,000 feet high was built. What was probably the last significant eruptive sequence in the Goat Rocks began about 600,000 years ago, when andesite from vents now marked by Ives Peak and Old Snowy Mountain flooded southeast into the headwaters of the Cispus River.

Because the interior of the Goat Rocks is so remote, it is difficult to reach some portions of the wilderness by round-trip day hikes. The following trips describe the shortest routes to points of interest. Because these

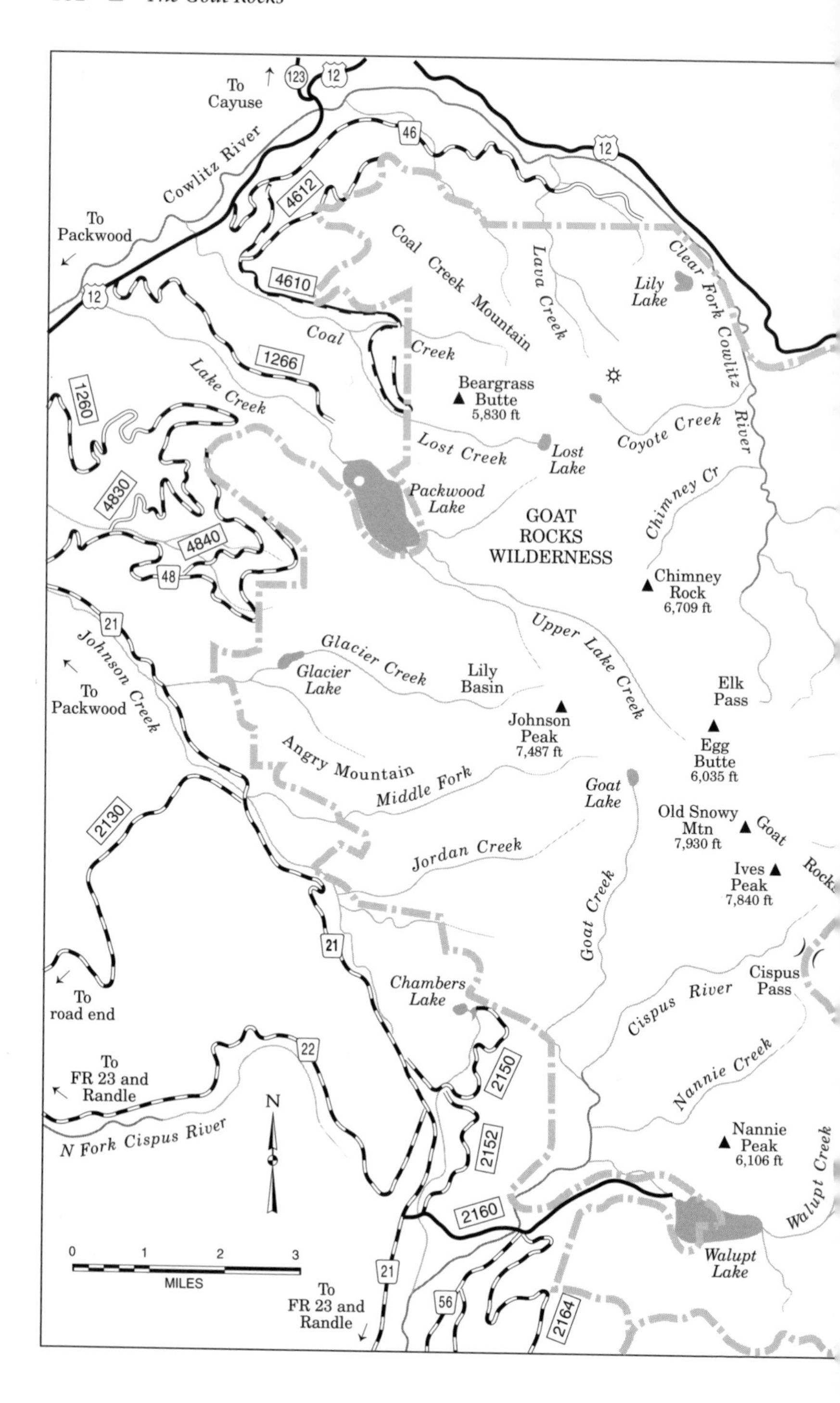

To Cayuse
123
12
46
12
Cowlitz River
4612
To Packwood
4610
12
Coal Creek Mountain
Lava Creek
Clear Fork Cowlitz River
Lily Lake
Coal
Creek
1266
Lake Creek
1260
Beargrass Butte 5,830 ft
Lost Creek
Lost Lake
Coyote Creek
4830
Packwood Lake
GOAT ROCKS WILDERNESS
Chimney Cr
4840
48
Chimney Rock 6,709 ft
21
Johnson Creek
To Packwood
Glacier Creek
Glacier Lake
Lily Basin
Upper Lake Creek
Elk Pass
Johnson Peak 7,487 ft
Egg Butte 6,035 ft
Angry Mountain
Middle Fork
Goat Lake
2130
Old Snowy Mtn 7,930 ft
Goat Rocks
Jordan Creek
Ives Peak 7,840 ft
Goat Creek
21
Chambers Lake
Cispus River
Cispus Pass
To road end
Nannie Creek
22
2150
To FR 23 and Randle
N
N Fork Cispus River
2152
Nannie Peak 6,106 ft
Walupt Creek
2160
Walupt Lake
0
1
2
3
MILES
21
To FR 23 and Randle
56
2164

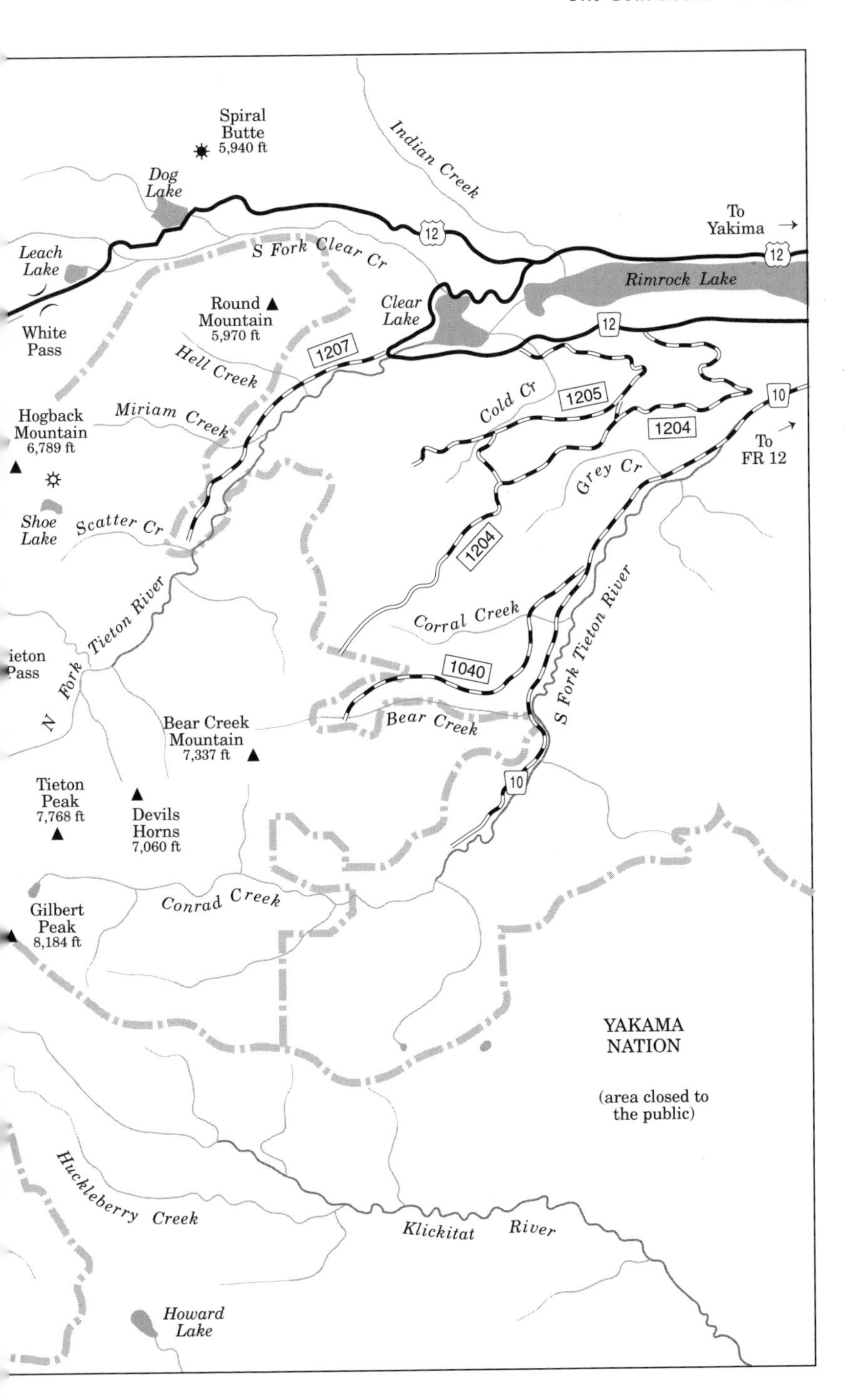
Spiral Butte 5,940 ft
Indian Creek
Dog Lake
To Yakima →
12
Leach Lake
S Fork Clear Cr
Rimrock Lake
Round Mountain 5,970 ft
Clear Lake
White Pass
1207
Hell Creek
Cold Cr
1205
Hogback Mountain 6,789 ft
Miriam Creek
1204
10
To FR 12
Grey Cr
Shoe Lake
Scatter Cr
1204
N Fork Tieton River
Corral Creek
S Fork Tieton River
ieton Pass
1040
Bear Creek Mountain 7,337 ft
Bear Creek
Tieton Peak 7,768 ft
Devils Horns 7,060 ft
Conrad Creek
Gilbert Peak 8,184 ft
YAKAMA NATION
(area closed to the public)
Huckleberry Creek
Klickitat River
Howard Lake

are the most popular hikes in the area, trails may be crowded. Some of these routes are long day hikes; an overnight camping trip may make them more enjoyable. An excursion the length of the wilderness via the PCT can link several of these features in a single multi-day trip.

Hogback Mountain

Hiking trail to volcano vents and flows

Trail: PCT #2000
Rating: (M); hikers, saddle and pack stock
Distance: Shoe Lake 7.2 miles
Elevation: Trailhead 4,620 feet, Hogback Mountain 6,789 feet, Shoe Lake 6,112 feet
Maps: USGS Old Snowy, Pinegrass Ridge, Spiral Butte, White Pass; USFS Packwood Ranger District
Driving Directions: Take Highway 12 to the trailhead at the entrance to Leech Lake Campground 0.5 mile east of White Pass.

Beneath the manicured ski slopes on the south side of White Pass lie basalt lava flows from another of the major volcanoes of the Goat Rocks Wilderness, Hogback Mountain. The summit of Hogback, which is visible on the horizon south from the top of the chairlift, tops the north–south ridge paralleling the route of the PCT from White Pass into the heart of the Goat Rocks Wilderness. Striking evidence of this past volcanic activity is visible from a saddle on PCT #2000, south of the pass.

The PCT leaves US 12 opposite the entrance road to Leech Lake Campground, 0.6 mile east of the pass, then makes a chain of a dozen easy switchbacks up densely wooded slopes to a flat containing tiny Ginnette

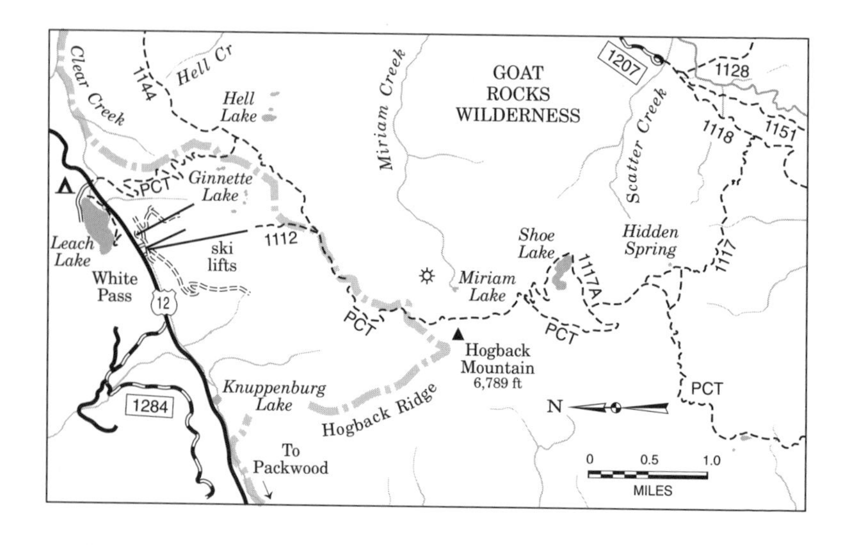

Lake. Just above the lake is the junction with Trail #1144 from the east; here the PCT turns southwest and ascends the broad ridge top to a wooded saddle. The path briefly crosses west of the ridge to avoid east face cliffs at the head of the Miriam Creek drainage. Once past this obstacle, the route recrosses the crest and begins a gradual diagonal ascent of the bare rock slopes high above the Miriam Lake basin. At this point, the summit of Hogback Mountain, 400 vertical feet above, is a relatively easy cross-country scramble southwest over loose rock.

The trail continues to traverse south across steep barren slopes below the snaggled crest until it reaches a saddle overlooking the glacier-carved cirque enclosing Shoe Lake. Either this saddle or the short ridge top to the east was the location of the vent that 2 million years ago poured forth the multiple, thin, viscous basalt flows that make up the Hogback Mountain shield volcano. The individual layers of the flow can be seen stacked like sheets of cardboard in the side slopes of the surrounding ridges. In spots where the lava flowed across snow or ice, the basalt takes on a more lumpy pillow-shaped characteristic of such contacts.

The old route of the PCT once switchbacked down past the shore of Shoe Lake, formerly a choice camping spot, then climbed southwest out of the basin. The popularity of the lake was more than the fragile ecology could cope with, and the surrounding meadow was rapidly being beaten to bare dirt. The area is now closed to camping, and the PCT has been rerouted along the top of the ridge to the west. The lakeshore trail is open to day travel; please treat the area gently so that it can survive for hikers of the future to enjoy.

Old Snowy Mountain and Ives Peak

Hiking trails and cross-country route to volcano vents

Trails: Snowgrass Trail #96, PCT #2000
Rating: #96 (M), #2000 (D); hikers, saddle and pack stock
Distance: 7.5 miles
Elevation: Trailhead 4,640 feet, Old Snowy summit 7,930 feet
Maps: USGS Hamilton Buttes, Old Snowy, Walupt Lake; USFS Packwood Ranger District
Driving Directions: 2.5 miles southwest of Packwood, turn south from US 12 onto FR 21 (2-lane gravel), and in 13.5 miles head east on FR 2150 (1-lane gravel). In 2.8 miles bear right on FR 2150040, and in 200 feet turn right again on FR 2150405 to the hiker trailhead for Trail #96 (the stock trailhead is 0.3 mile farther on FR 2150040).

This is probably the shortest trail into the alpine country on the southwest side of the Goat Rocks so, although it is loaded with spectacular scenery, don't expect a lot of solitude. To reduce trailhead congestion and accommodate their particular needs, horse-borne and foot-borne travelers have separate start points for Trail #96, but the paths merge in about ¼ mile.

From there the trail makes a gradual side-slope descent into the bottom of the deep glacier-carved valley drained by Goat Creek.

A hasty crossing of the soggy, mosquito-filled bottomland adjoining the creek leads to the base of the wooded wall on the east side of the drainage. The trail steeply diagonals up this side slope; 800 feet higher and a few switchbacks later, it reaches more moderate slopes where the forest cover begins to thin. From here the trail once followed a tributary of the Cispus River straight uphill through the fabulous flower meadows of Snowgrass Flat. Too many hooves, grazing teeth, boots, tent sites, and fire rings were reducing this sublime spot to an alpine dustbowl, and in 1980 a thick layer of ash from the St. Helens eruption nearly administered the coup de grace. To help the fragile environment recover, Trail #96 has been rerouted to a ridgeline to the west, and camping is prohibited in the meadow. Please treat this area with care and help it recover its former beauty.

The new path follows the lightly timbered ridge north as it climbs past a junction with Trail #86 from Goat Lake, then ends as it reaches the PCT in alpine parkland at 6,420 feet. In early summer much of the trail north from here may be snow-covered, but later in the season it makes a rapid transition from meadow to glacial till, talus, and blocks of glacier-polished andesite. This is not the place to be caught by bad weather, because there

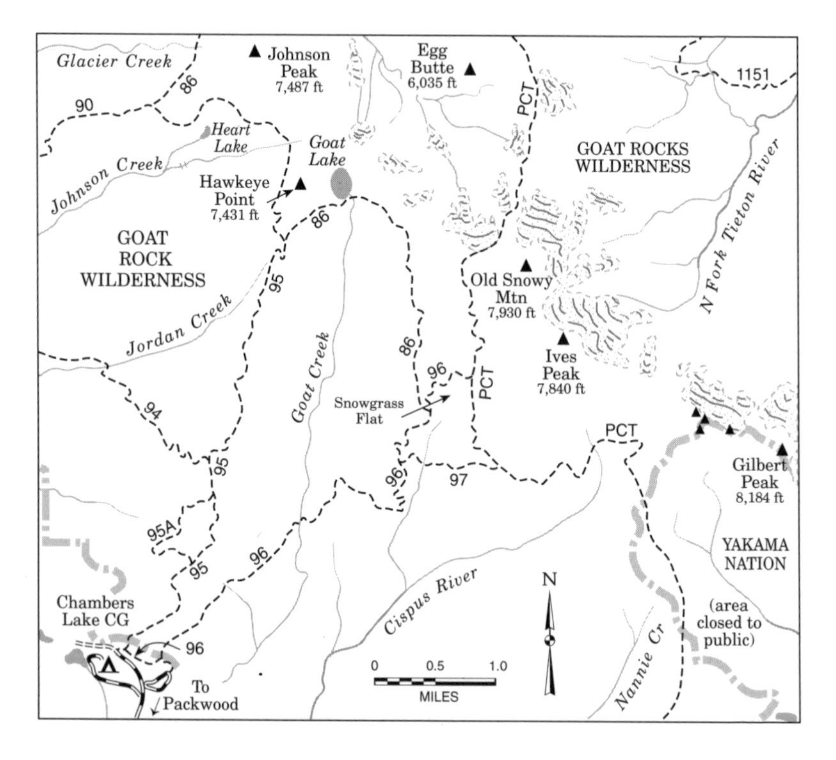

is scant protection from wind, rain, and snow. Storm-caused deaths have occurred in this section of the PCT.

The ridge above, capped by the summits of Old Snowy, Ives Peak, and lesser summits between the two, was built by younger volcanic dikes and vents that extruded at least four separate floods of andesite lava from the Goat Rocks backbone sometime between 600,000 and 140,000 years ago, long after the bulk of the older Goat Rocks Volcano had eroded away. These lava flows built up the alpine plateau between the present drainages of Goat Creek and the Cispus River and continued downstream to near the present-day location of FR 21. The deep bordering valleys were cut later by advances of alpine glaciers.

As the PCT weaves uphill north toward the west shoulder of Old Snowy, views overlook the ragged ridge crest extending west above the glacial cirque embracing Goat Lake and on beyond to the summit of Johnson Peak. This long rib and the intrusive andesite dikes displayed in its walls are remnants of other vent sites of the Goat Rocks Volcano complex. When the route reaches the ridge west from Old Snowy's summit, it also reaches the highest point of the PCT in the state of Washington, 7,080 feet. The original PCT went higher yet, climbing east on this ridge for another 500 feet before heading north along the very top of the narrow crest rib. By taking this old abandoned route to the 7,600-foot level on the ridge, the 7,930-foot-high summit of Old Snowy is just an 800-foot scramble away.

Southeast, across the deep headwater valley of the North Fork of the Tieton are Gilbert Peak, Goat Citadel, Black Thumb, and Big Horn, all black pinnacle remnants of one vent area of the ancient Goat Rocks Volcano. To the northwest, towering above the equally deep Upper Lake Creek drainage, is the long Johnson Ridge massif, another vent complex for that same volcano.

Bear Creek Mountain

Hiking trail to views of volcano vents

Trails: Bear Creek Mountain Trail #1130, Bear Creek Lookout Trail #1130A
Rating: (M), hikers, saddle and pack stock
Distance: 3.6 miles
Elevation: Trailhead 6,030 feet, summit 7,337 feet
Maps: USGS Old Snowy, Pinegrass Ridge; USFS Naches Ranger District
Driving Directions: From US 12, 8 miles east of White Pass, turn south on Tieton Road (2-lane paved), signed to Clear Lake. In 7 miles turn south on FR 1204 (1-lane gravel), and follow it for 7.9 miles to the junction with FR 1204760. Continue southwest on FR 1204 (rough, narrow, 1-lane dirt) for another 2.6 miles to the road end at Section 3 Lake.

Bear Creek Mountain sits at the upper end of a high lava plateau between the North and South forks of the Tieton River. Its 7,337-foot-high summit, only 847 feet lower than the highest spot in the Goat Rocks

The view from Bear Creek Mountain. Devils Horns are the rocky pinnacles in the right foreground, and Devils Washbasin is right of the peak, behind the ridge. Tieton Peak is the tall mountain on the far right, and Gilbert Peak is in the distance.

Wilderness, commands sweeping views of the broad backbone of the Goat Rocks, a scant 5 miles to the southwest. To the north it overlooks the high ridges and deep valleys of the White Pass area and the William O. Douglas Wilderness.

The approach road deteriorates to a single-lane, potholed, exposed, dirt path as it nears the road-end turnaround at tiny, man-made Section 3 Lake (inventively named because it lies in Section 3 of its range and township).

The trail as far as the summit spur is 2.5 miles of pleasure, more than compensating for the road. After starting out as a moderate ascent through open forest of Pacific silver fir, subalpine fir, and mountain hemlock, the way crosses three successive alpine meadows, broad greenswards that sweep up to short talus slopes below ridge-top rock outcrops. Depending on the season, hellebore, lupine, or purple gentian frame the trailside. Both meadows and trail are covered by ¼ inch or more of fine light gray powder, an ash legacy of the 1980 eruption of St. Helens. The third meadow provides the initial view ahead of the talus-and-cliff slopes that fall away from the summit of Bear Creek Mountain. After crossing a few small creeks and a

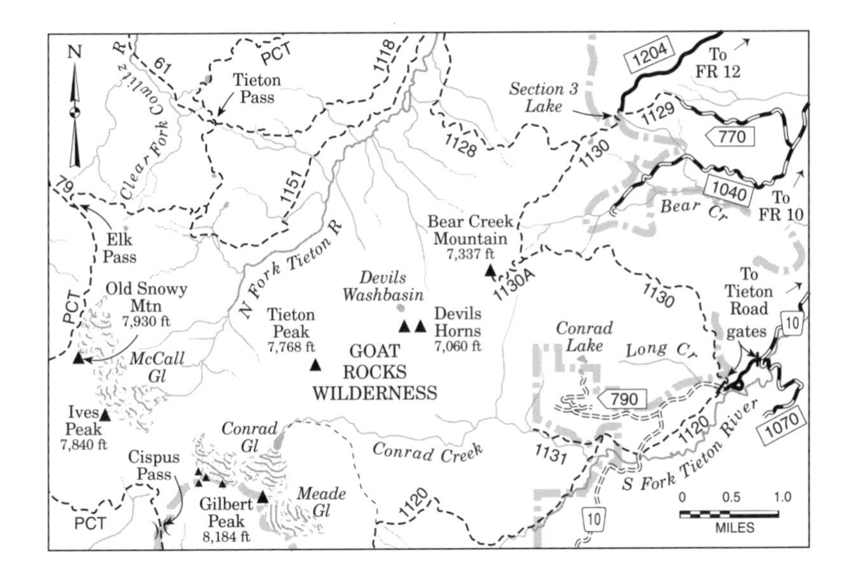

steep pumice sidehill, the path reaches Trail #1130A, the spur to the old lookout site at the summit.

Savor memories of the easy meadow hike, it's now over—the summit trail is not maintained and was nothing great even when it was. The unmarked path is generally easy to follow through an endless succession of switchbacks, but it is very steep, and the footing is on a loose rubble of tephra, pumice, and fragments of old lava flows. It weaves upward past exposed dikes and cliff-face layers of platy andesite and bulbous basalt before reaching the ridge crest. The views from here are great, and they get better as the path wends another 150 feet higher to the cliff-bound summit platform. The fire lookout once located here was staffed between the 1930s and 1945, but the cabin was destroyed in the 1960s. But, oh, the marvelous views remain!

A narrow knife-edge ridge leads southeast to the aptly named brick-red pinnacles of Devils Horns. These basalt towers mark the site of a small 3-million-year-old volcano, whose vent was centered at Devils Washbasin, a tarn embraced by the ridges falling away north from the Horns. The lighter yellow-gray base below Devils Horns outcrops is a remnant of a slightly older eruption of rhyolite from a vent in the same area, one of the earliest examples of volcanic activity here in the last 4 million years.

Beyond Devils Horns the ridge continues southwest to Tieton Peak. It and another unnamed summit to its south are banded with a series of narrow horizontal cliffs that record layers of andesite lava flows over a period of 2 million years. These flows were fed by vents at the core of the Goat Rocks Volcano, vestiges of which make up the black pinnacles of Big Horn, Little Horn, Goat Citadel, and Gilbert Peak along the skyline ridge to the southwest. Had you been sitting at this same spot a million years ago, you would probably have been looking at a scene much like what you would see today at Paradise on Mount Rainier, with the Goat Rocks Volcano substituted for Rainier.

Cispus Pass

Hiking trail to views of volcano vents

Trails: Nannie Ridge Trail #98, PCT #2000
Rating: (M); hikers, saddle and pack stock
Distance: 6.5 miles
Elevation: Walupt Lake trailhead 3,196 feet, Cispus Pass 6,460 feet
Maps: USGS Walupt Lake, USFS Packwood Ranger District
Driving Directions: 2.5 miles southwest of Packwood, turn south from US 12 onto FR 21 (2-lane gravel). In 16.6 miles head east on FR 2160 (1-lane paved), and follow it for 3.3 miles to Walupt Lake Campground.

This trip includes a scenic ridge-top hike at timberline offering wide panoramic vistas of the south end of the Goat Rocks Wilderness, followed by a section of the PCT with "reach out and touch" views of the high ragged pinnacles at the southeast end of the Goat Rocks backbone ridge. It all starts at Walupt Lake Campground; from here take Trail #101 a few hundred yards along the north shore to where Trail #98 heads uphill to the north. A gradual climb through dense forest becomes progressively steeper, finally reaching the point where switchbacks ease the grade. Soon even these become steep and more frequent until at last, after a gain of 1,900 feet of elevation, trees thin and the path reaches a ridge-top saddle.

From the saddle hikers with an experienced eye can pick out an

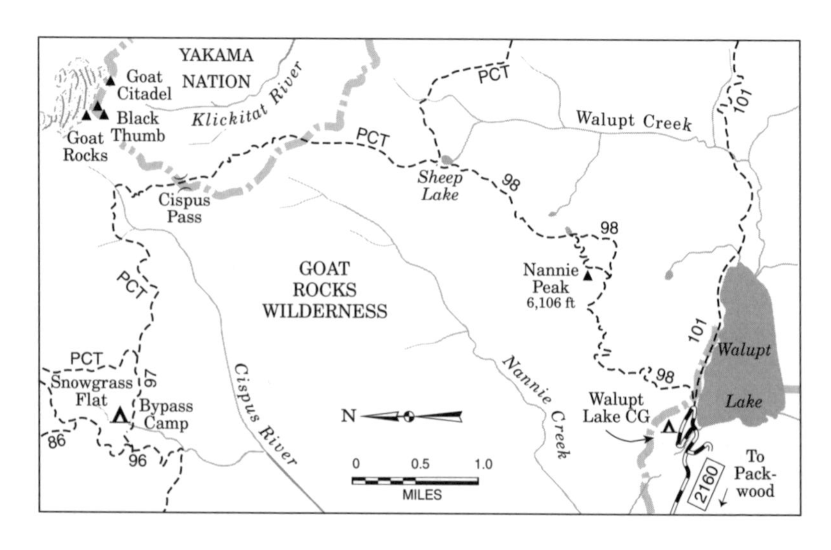

Cispus Pass lies atop a massive, million-year-old andesite intrusion.

abandoned path north up the crest to the top of Nannie Peak, an old lookout site. The main trail drops off the ridge to traverse below cliffs on the east face of the peak, then regains the ridge top about ¾ mile to the northeast. The path traverses the open southeast rim of the ridge with continuous views across the wide, deep Walupt Creek drainage to the Cascade crest and south to the broad, forested basalt shield flowing north from the vent of the Lakeview Mountain volcano.

A flat below a step in the ridge harbors pretty, meadow-wrapped Sheep Lake. Just above the lake Trail #98 ends at its junction with the PCT. The old PCT traversed around the head of Nannie Creek and the rock nose separating it from the Cispus River drainage to regain the crest just north of Cispus Pass. However, this section of the trail has been relocated to add a little spice to the route; instead of a traverse from the junction with Trail #98, it now diagonals up a steep sidehill to a saddle, where it crosses over into the headwaters of the Klickitat River and the Yakama Indian Reservation.

For the next 1.2 miles the ribbon-narrow path traverses steep, rocky slopes with views southeast down the cavernous glacier-carved trough of the upper Klickitat River. The opposite wall of the canyon rises abruptly to a sheer 1,000-foot-high cliff capped by the jagged lava plug pinnacles of Black Thumb, Little Horn, Big Horn, Goat Citadel, and the highest spot in the Goat Rocks, Gilbert Peak. The gap at the end of the basin headwall is Cispus Pass, your destination.

The pass and the rough horseshoe ridge to its west at the headwaters of the Cispus River lie atop the 1-million-year-old Cispus Pass pluton, a massive andesite intrusion thought by many to mark the magma source that created the Goat Rocks Volcano.

The Clear Fork Lava Flow and Coal Creek Mountain

Hiking trail past a volcano vent and dikes

Trail: Clear Lost Trail #76
Rating:(D); hikers, saddle and pack stock
Distance: 6.6 miles
Elevation: Trailhead 3,560 feet, Clear Fork of the Cowlitz River crossing 3,150 feet, Coal Creek Mountain 6,376 feet
Maps: USGS Old Snowy, Packwood Lake, White Pass; USFS Packwood Ranger District
Driving Directions: The trailhead is on the south side of US 12, 3 miles west of White Pass.

Solitude is often proportionate to difficulty, thus you probably will meet few other hikers on Trail #76 between US 12 and the southeast tip of Coal Creek Mountain. Two other less difficult trails reach the summit: Trail #65 up Coal Creek Mountain from the northwest and Trail #78 up the west side slope from Packwood Lake. Those trails, however, do not pass the stunning geological features seen on Trail #76. Hikers may choose to

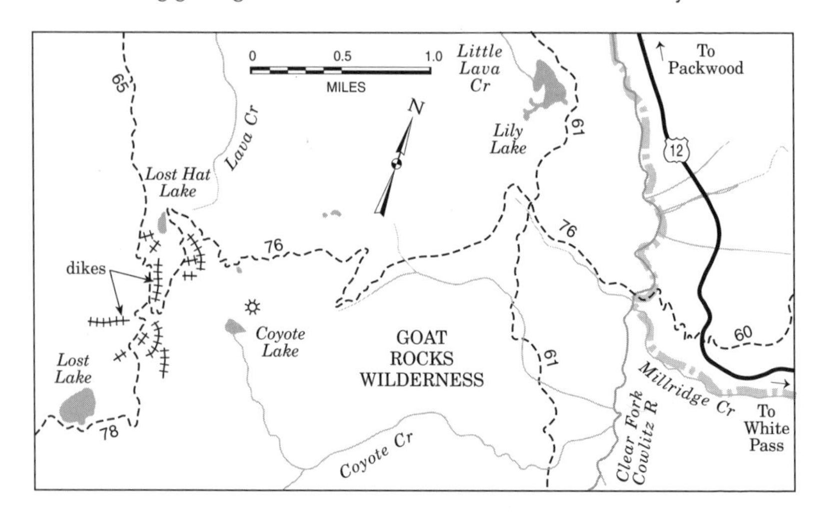

take either of the two alternate trails for the return trip; however, road-end transportation must be prearranged because both are more than 20 road miles from the US 12 start point of Trail #76.

At the start, Trail #76 commits an unpardonable sin; it loses 400 feet of elevation en route to a target over 2,800 feet higher than the trailhead. After this transgression, the path reaches a crossing of the Clear Fork of the Cowlitz River—but no bridge. The only solution is a cold, difficult ford of the river.

The elevation lost reaching the river is regained as the tread heads up the edge of the massive Clear Fork lava flow to its flat bluff-like top, now overlain by a thick layer of glacial till. Here the trail crosses Clear Fork Trail #61 then resumes its climb up the steep, thickly wooded side slope toward the source of this 100,000-year-old dacite eruption, the ridge top north of Coyote Lake.

At 5,100 feet the path reaches a saddle between the source vent ridge and a long high crest that splits the Lava Creek and Clear Fork drainages. This same crest also split the Clear Fork dacite flow and stood as an island surrounded by the lava that flooded down channels on either side. As the trail continues its ascent west, it briefly breaks out of woods to spots where shelf-like outcrops of the lava flow can be spotted.

After a brief respite at a small tarn wrapped in the cliff walls of a glacial cirque, the path once more climbs across a talus and boulder hillside to yet another cirque. Here the narrow, rock-encircled finger of Lost Hat Lake nestles below 250-foot-high dike-cut cliffs on the northeast face of Coal Creek Mountain. Because of its altitude and north face location, the lake remains frozen most of the year. The trail leaves the lake to climb once more, this time up a bordering ridge, then crosses a bare rock side slope that offers good views east to the bald knob that marks the vent source of the Clear Fork flow. A few switchbacks gain the high point at the south end of Coal Creek Mountain, formerly the site of a fire lookout. The tent cabin placed here in 1922 was replaced with a more permanent structure in 1934; it was later abandoned, and all signs of it disappeared by the 1960s.

The bare flanks of this end of Coal Creek Mountain are laced with a swarm of radial andesite dikes that intruded through weak spots in the overlying lava formations some 35 to 28 million years ago. These most likely indicate an underlying shallow pluton that pressed up beneath the summit in the same period.

12

The White Pass Corridor

From a geological standpoint, the area surrounding US 12 from its junction with Highway 123 northeast of Packwood to the Wenatchee National Forest boundary east of White Pass is the most unique section of the South Cascades. Here, within a 30-mile-long corridor, are more than 150 million years of geological history of southwest Washington displayed in roadcuts through ancient rock formations, side walls of deep stream-carved valleys, lakes created by lava flow dams, cone-shaped volcano domes, faults and folds in base rock structures, and the sheer cliff walls of magma intrusions. This narrow east–west strip shows a more complete sample of South Cascades volcanic activity than anywhere else in the South Cascades. With a few exceptions farther north in the William O. Douglas Wilderness, the White Pass corridor appears to mark the northernmost extent of major volcanic activity in the South Cascades within the last 10 million years.

US 12 Roadcuts

Road trip past layered rock formations

Elevation: 1,600 to 4,470 feet
Maps: USGS Old Snowy Mountain, Rimrock Lake, Spiral Butte, Tieton Basin, White Pass; USFS Packwood and Naches ranger districts
Driving Directions: From the west, leave I-5 at Exit 68, and head east on US 12 for 74.5 miles to the junction with Highway 123, the west end of the White Pass corridor. From the north, take Highway 410 southeast from Enumclaw for 72 miles to Cayuse Pass, then head south on Highway 123 for 16.9 miles to its junction with US 12. From the east, take US 12 west from Yakima for 30 miles to the east end of the White Pass corridor.

By necessity, the highway engineers who laid out the route of US 12 over White Pass were forced to make many deep roadcuts into the steep side slopes of the drainages leading up to the pass. In the process, they exposed numerous fascinating examples of millions of years of accumulated sedimentary and metamorphic rocks, lava flows, and volcanic debris that make up the geological structure of the White Pass corridor. The unfortunate part of this highway construction is that much of this rock is a bit unstable and has an unnerving habit of peeling off sidehills and

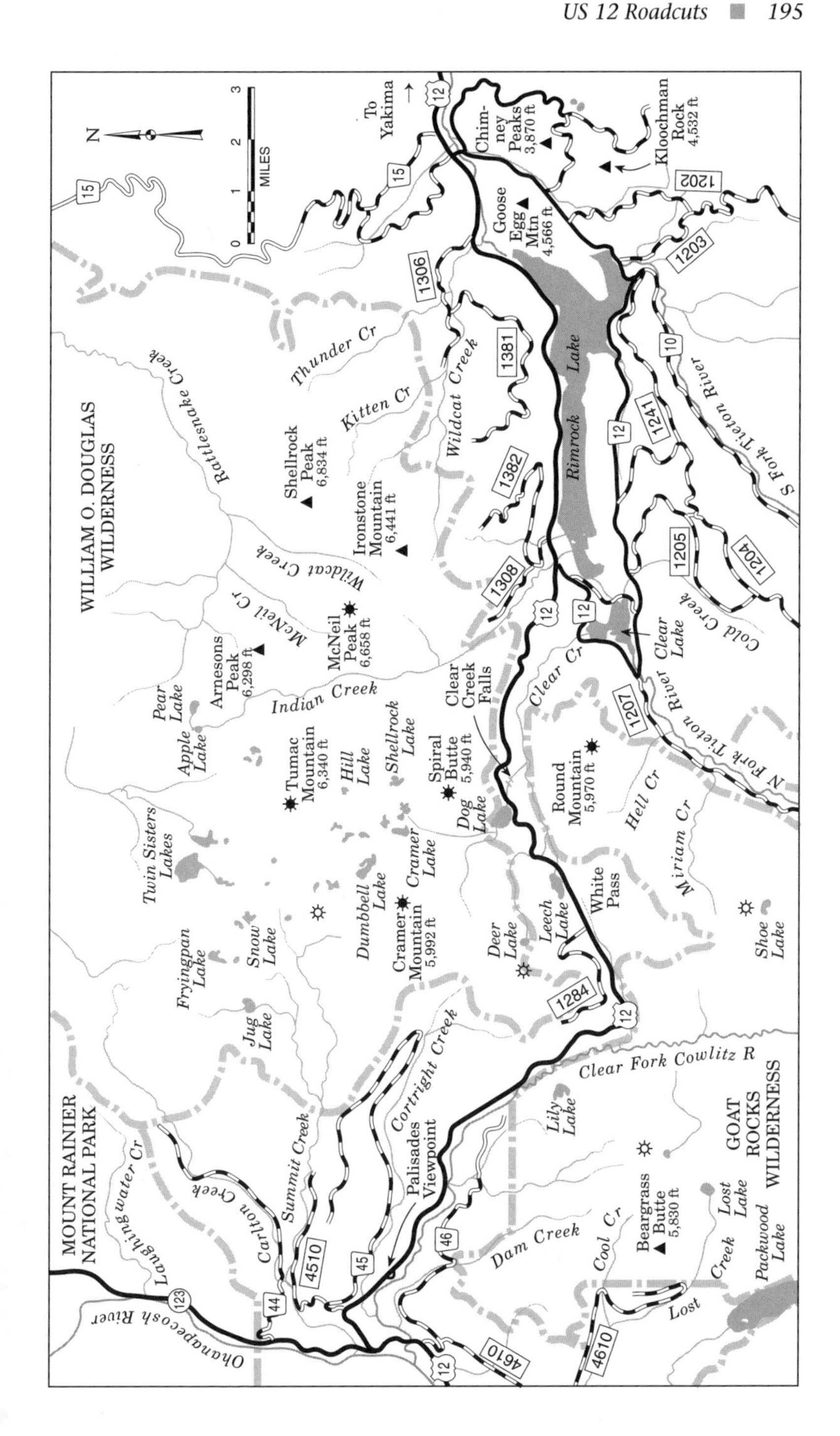
WILLIAM O. DOUGLAS WILDERNESS
MOUNT RAINIER NATIONAL PARK
GOAT ROCKS WILDERNESS
N
0 1 2 3
MILES
To Yakima
Chimney Peaks 3,870 ft
Kloochman Rock 4,532 ft
Goose Egg Mtn 4,566 ft
Rimrock Lake
Clear Lake
Shellrock Peak 6,834 ft
Ironstone Mountain 6,441 ft
McNeil Peak 6,658 ft
Arnesons Peak 6,298 ft
Tumac Mountain 6,340 ft
Spiral Butte 5,940 ft
Round Mountain 5,970 ft
Cramer Mountain 5,992 ft
Beargrass Butte 5,830 ft
Clear Creek Falls
White Pass
Palisades Viewpoint
Thunder Cr
Kitten Cr
Wildcat Creek
Rattlesnake Creek
McNeil Cr
Indian Creek
Clear Cr
Cold Creek
S Fork Tieton River
N Fork Tieton River
Hell Cr
Miriam Cr
Pear Lake
Apple Lake
Hill Lake
Shellrock Lake
Dog Lake
Twin Sisters Lakes
Dumbbell Lake
Cramer Lake
Fryingpan Lake
Snow Lake
Jug Lake
Deer Lake
Leech Lake
Shoe Lake
Lily Lake
Lost Lake
Packwood Lake
Lost Creek
Cool Cr
Dam Creek
Clear Fork Cowlitz R
Cortright Creek
Summit Creek
Carlton Creek
Laughingwater Cr
Ohanapecosh River
12
15
123
10
44
45
46
1202
1203
1204
1205
1207
1241
1284
1306
1308
1381
1382
4510
4610

rolling across the highway at inopportune times. As a result, some of the best spots for observing the form and composition of the roadcuts are signed "No Parking or Standing"; pulling off for more than a fleeting glance may earn you a lecture from the State Patrol, possibly a ticket, or a boulder in your lap! For those who wish to sightsee, pick a designated driver who is bored by scenery and geology, and exploit a passenger's opportunity to rubberneck at surrounding hillsides.

Following are brief descriptions of some of the most interesting of the US 12 roadcuts in the White Pass corridor, starting from the west end at the junction with Highway 123 and proceeding to the east.

Rock outcrops in roadcuts for the first 7 miles were primarily created by the earliest stage of volcanic eruptions in the White Pass area. They consist of multiple layers of ejected andesite and dacite fragments, andesite lava flows, and other volcanic debris. The eruptions that created them took place in a lowland trough but added little to the area's net elevation because the trough slowly subsided under the weight of the newly accumulated volcanic material. The stage of volcanism that created this formation started about 38 million years ago and continued for about 10 million years, building an accumulation of volcanic rocks between 7,500 and 10,000 feet thick.

A particularly striking example of the strata within this formation is found 2 miles east of Highway 123 at Cortright Creek. Pulling off onto the

Tilted rock strata seen along US 12 show the "basement" formations of the South Cascades.

shoulder here is permitted. Examine the exposed rock face on the northwest side of the creek, where multiple layers of these volcanic flows and blocky debris diagonal steeply up the drainage wall. At 2.3 miles is the Palisades Viewpoint, described in full later.

A bad slide area 7 miles from Highway 123 marks the contact point between this accumulation of volcanic debris (on the west) and a much older underlying formation that is exposed in many roadcuts from here east to Rimrock Lake. This older rock, the basement beneath the southern Washington Cascades, is a melange of volcanic sediments that had originally accumulated on the ocean floor off the west coast of North America, perhaps 140 to 120 million years ago. When the oceanic plate beneath this rock advanced northwest and subducted beneath the continental crust, the surface of the plate was scraped off, kneaded, and deformed to create the lowland upon which the Cascade volcanoes were later formed. Starting about 17 million years ago, magma that had accumulated beneath this formation forced it upward, then the overlying formations were eroded from resulting Cascade backbone. The White Pass area is the only window in southern Washington where the "roof" has been removed from this ancient rock and the old "basement" of the Cascades is exposed.

At White Pass, three different geological formations are represented in ski slopes to the south. The lowest is the ancient rock described in the preceding paragraph. About 500 feet upslope are outcrops of residual lava from the Goat Rocks Volcano, whose origin was described in chapter 11. Cliffs below the top of the main chairlift mark the lower end of layers of basalt lava that erupted about 2 million years ago from the Hogback Mountain shield volcano 3 miles to the south. On the north side of the highway are andesite flows from a more recent volcano vent, 650,000-year-old Deer Lake Mountain volcano, just south of Deer Lake. This vent extruded lava to the south and west that blocked creek drainages to create Leech Lake, where White Pass Campground is now located.

Immediately east of White Pass the obvious volcano cone of Spiral Butte rises on the north side of the road. About 3 miles east of the pass the road slices through the lower end of the Spiral Butte dacite flows. In roadcuts on the north side, these can be recognized as the many thin, downsloping sheets of rock that, by virtue of their fracture planes and slope angle, are a continuous rock slide problem threatening the highway.

About 5½ miles east of the pass are two distinct fault lines, a little over a mile apart. Between these faults is the oldest recognizable rock in the White Pass vicinity, about 145 million years old. Like the basement rock seen in roadcuts to the west, this rock was also sheared from the subducting ocean floor, buried, and then churned upward by collision of the North American and oceanic plates. It was most probably the core of a seamount, a volcano that erupted and built up on the ocean floor. The heat and pressures resulting from its attachment to the continent have twice restructured (metamorphosed) the original rock. The faults on both sides of this block mark the "suture" lines where this older rock and younger sediments along its edges were joined to the continent. Several near-vertical dikes can be seen intruding the formation between the two fault lines.

Near the east end of Rimrock Lake, 13 miles east of the pass, the rock on the north side of the road shifts from ancient sandstones that were the base formation at the pass to younger Westfall Rocks. This mass of diorite that intruded overlying strata about 25 million years ago is just one of many such local surface exposures that may have been spawned from a single underlying pluton.

The Palisades

Roadside viewpoint of a lava flow

Elevation: 2,200 feet
Maps: USGS Ohanapecosh Hot Springs, USFS Packwood Ranger District
Driving Directions: Take US 12 east from the junction with Highway 123 for 2.3 miles to the turnoff on the south side of the road at the Palisades Viewpoint.

What is probably one of the area's most accessible and dramatic examples of columnar jointing in lava flows can be seen at a roadside interpretive site on US 12, 2.3 miles east of its junction with Highway 123. Here, as seen at a grassy overlook, the Clear Fork of the Cowlitz River has cut a 450-foot-deep canyon through a melange of relatively soft rock created mostly of volcanic debris. The river cut exposes a harder, 200-foot-thick, columnar-jointed, dacite lava flow in the opposite wall of the canyon. This lava was extruded sometime between 110,000 and 20,000 years ago from a volcano vent high on the ridge above Coyote Lake, about 9 miles upstream to the southeast. (See The Clear Fork Lava Flow and Coal Creek Mountain in chapter 11.)

Columnar jointing is vividly displayed in this wall at a viewpoint along US 12.

The flow is flat and uncommonly thick at this point, indicating that it must have ponded against some type of obstruction a short distance to the west. This has been theorized to be either a scarp (a wall created by the upthrust side of a fault) or possibly a lobe of an alpine glacier that may have extended into the drainage about the time of the eruption. The spectacular columnar joints extend from the top to the bottom of the dacite wall, perpendicular to the direction of the lava flow. Such jointing is caused when lava cools rapidly and shrinks along the axis of cooling, fracturing the rock into long columns parallel to that axis.

Other, lesser examples of columnar jointing can be seen farther upstream in the Clear Fork drainage and in a few places in the lower Lava Creek drainage. The lava flow split just below its source vent, and portions flooded down both sides of a high, elongated ridge between Lava Creek and the Clear Fork, then merged again in the lower Lava Creek drainage.

Spiral Butte and Deer Lake Mountain

Hiking trail to a volcano vent

Trails: Sand Ridge Trail #1104, Shellrock Lake Trail #1142, Big Peak Trail #1108
Rating: #1104 and #1142 (M), #1108 (D); hikers, saddle and pack stock, #1108 not recommended for stock
Distance: 6 miles
Elevation: Trailhead 3,400 feet, summit 5,940 feet
Maps: USGS Spiral Butte, White Pass; USFS Naches Ranger District
Driving Directions: 6.2 miles east of White Pass turn north from US 12 onto FR 1200488 to reach the start of Trail #1104 in 0.2 mile.

Two young, very unique volcanoes just north of White Pass were key in creating many of the popular present-day recreational and interpretive features in the vicinity of the summit. One of these just east of the pass, Spiral Butte, is a readily identifiable volcano cone; the other immediately west of the pass, Deer Lake Mountain, is heavily wooded on all but the west face and is not as easily recognized as a volcano vent. Both vents were active sometime between 110,000 and 20,000 years ago, although Spiral Butte is the younger. A hiking trail leads to the summit of Spiral Butte, and the PCT crosses the Deer Lake Mountain lava flow east of its summit.

From US 12 at White Pass, heavily wooded Deer Lake Mountain rises about 820 feet above the surrounding terrain. Its shape resembles a volcano cone only when viewed from FR 1284, which skirts the base of a steep, bare talus slope on the west side of its summit. Andesite lava flooded west from the summit crater for nearly 3½ miles to the west shore of Dog Lake; two smaller lobes of the flow twisted south, blocking stream drainages to frame and form Leech Lake, the site of today's White Pass Campground. Some platy rock layers of the flow can be seen in the open hillside at the west end of the lake.

Spiral Butte rises 1,800 feet above the east shore of Dog Lake, just north of US 12. Barren talus slopes that drop abruptly to the lake outline the flat-topped, inverted V-shape of the volcano cone. When viewed from the air, the contours of the mountain form a spiral beginning from the north side of the summit and twisting around its north and east sides in an ever-increasing arc—thus the name. (Until US 12 was completed across the pass, the peak suffered the more mundane name of Big Mountain.)

The bizarre flow pattern was created when the thick, viscous dacite that built the initial dome pooled up in the summit crater then breached the north side of its rubble rim. The lava drained down between channels formed by either volcanic rubble or glacial ice that encircled the vent; between thirty-five and forty-five individual thin lava flows spread down these channels to form the multiple layers of platy rock seen in the US 12 roadcuts east of the butte. The lava twisted far enough around the south side of the mountain to override the end of the Deer Lake flow and block Clear Creek to create Dog Lake.

In the 1930s a trail was built to the top of the butte as part of the Forest Service's plan to place a fire lookout on the summit. The project was dropped before the lookout was constructed, but the rugged seldom-maintained trail still exists. The shortest approach starts from FR 1200488, where Trail #1104 begins a gradual but steady ascent along the forested ridge to the northwest. In 3 miles Trail #1142 leaves to the west. Follow this trail as it bends northwest along the base of the rocky margins of

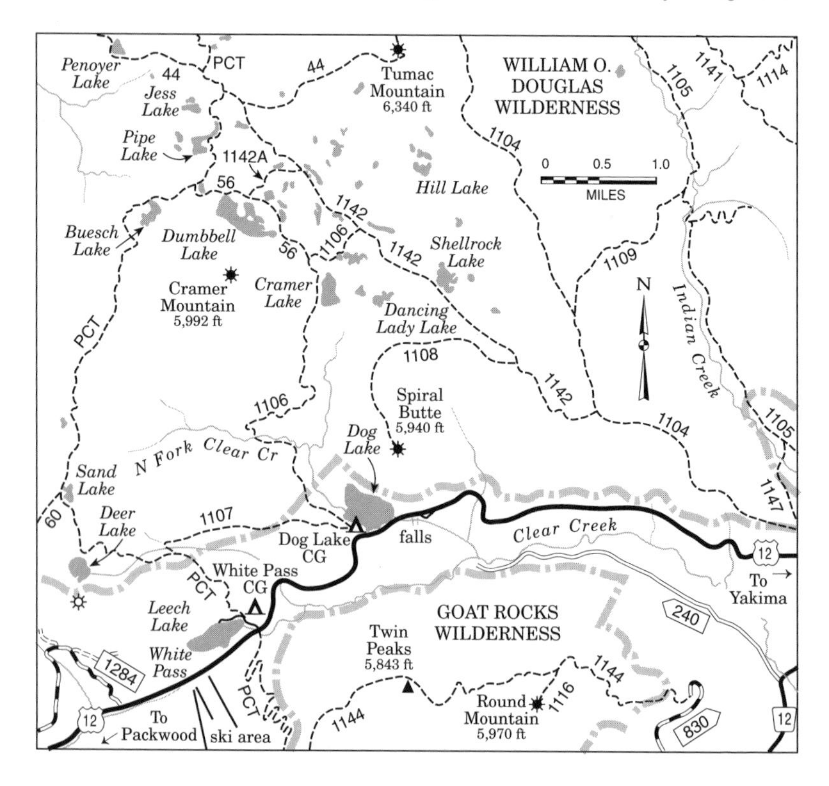

Spiral Butte is a dramatic cinder cone rising above Dog Lake.

the Spiral Butte lava flows. In another 0.8 mile, continue on Summit Trail #1108.

This path initially climbs a steep, wooded draw between two ribs bounding one of the channels of the lava flow, then diagonals up the south channel wall to the edge of the main spiral of dacite. The trail traces the margin of this flow as it circles west, then south through subalpine fir and mountain hemlock to the rock-ribbed summit. Outcrops here are the only remaining traces of the old volcanic core. Below, to the southwest, are the blue-green waters of Dog Lake and the ribbon of US 12 winding up to the ski slopes of White Pass. Due west is the forest-clad flow from the top of Deer Creek Mountain. North, the cinder cone of Tumac Mountain oversees the woods, meadows, lakes, and ponds of the broad plateau at the heart of the William O. Douglas Wilderness. The southern horizon is serrated by the peaks and pinnacles along the Goat Rocks backbone.

Clear Creek Falls

Roadside waterfall viewpoint

Elevation: 4,040 feet
Maps: USGS Spiral Butte, USFS Naches Ranger District
Driving Directions: From US 12, 2.6 miles east of White Pass, pull off south into the Clear Creek Falls Scenic Viewpoint.

Less than ¼ mile after leaving Dog Lake, Clear Creek drops over two of the most impressive waterfalls to be seen in the South Cascades. At the upper falls, the stream boils down a steep 50-foot-high lava ramp in a thick, narrow torrent, then spreads out to a fan of rivulets in its final, 25-foot plunge off the broadened toe of the chute. A few hundred yards downstream the stream plunges over a lip of rock in a 300-foot-high veil whose folds snag on mid-face outcrops, mimicking the form of the main cascade in a chain of smaller falls.

The walls that frame the falls, and the erosion-resistant lips over which they drop, are formed of a hard, 500-foot-deep stack of thin andesite plates, the lower margin of the lava flows that spread east from Deer Lake Mountain volcano. The creek cut through the underlying base of a soft ancient melange of rock below the falls as it flowed down the broad, glacier-carved valley to the east.

Tumac Mountain

Hiking trail to a volcano cone

Trails: Cramer Lake Trail #1106, Shellrock Lake Trail #1142, Cowlitz Trail #44
Rating: (M); hikers, saddle and pack stock
Distance: 9 miles
Elevation: Trailhead 4,240 feet, summit 6,340 feet
Maps: USGS Spiral Butte, White Pass; USFS Naches Ranger District
Driving Directions: Take US 12 2.1 miles east from White Pass to Dog Lake Campground and the start of Trail #1106.

The broad, high Tumac plateau north of White Pass lies deep in the heart of the William O. Douglas Wilderness. Over 200 lakes, ponds, and potholes, many surrounded by pretty little open meadows, lace the wooded plateau. Although early summer wildflower displays are outstanding, they are accompanied by hordes of mosquitoes, as one might expect of such a moist area. Many hikers prefer to defer their visit until fall, when most of the pesky critters are gone. The centerpiece of the plateau is Tumac Mountain, a distinctive cinder cone that hovers about 550 feet above a crater depression harboring a tiny lake.

Although the shortest route to the mountain comes in from the north from Bumping Lake, a far more interesting route, geologically, leaves Highway 12 east of White Pass at Dog Lake Campground. Fill out a wilderness permit at the start of Trail #1106, then head northwest on a gentle upward grade through dense hemlock and Douglas-fir. As you enter the wilderness in about ¼ mile, the forest becomes good-sized old-growth. At about 1 mile the path approaches then parallels Clear Creek. Its burbles and murmurs accompany the hike for another 0.5 mile to a footlog crossing. The route now bends east and starts a steep diagonal ascent; midway up the slope it traverses an open slide of andesite plates and boulders with a good view of the bald south face of the Spiral Butte volcano cone.

In another 0.5 mile the steepest section of trail is past, and the route

Tumac Mountain, which rises above the Tumac Plateau, is a relatively young volcano.

settles to an easy-grade ascent through open forest with an understory of vanilla leaf and beargrass. At about 3 miles is a marshy meadow, fringed with cottongrass. A spur leads a hundred yards east to Cramer Lake, a very popular camping and fishing site. Trail #1106 bends around the edge of the meadow and passes the intersection with Trail #56 to Dumbbell Lake. The route now heads east and descends to trace the wooded shore of an unnamed lake and then twist around the convoluted shoreline of its companion, Otter Lake. The path bends northeast, passes more small ponds, and crosses picturesque meadows to reach the junction with Trail #1142 from Shellrock Lake.

Head northwest on Trail #1142 as the near-flat track twists past several more ponds and a pair of tiny unnamed lakes. Pass the junction with Trail #1142A in a small meadow with still more minuscule lakes, then continue across the plateau past Benchmark Lake to the intersection with Trail #44. Here turn east and start a gradual ascent of the Tumac Mountain shield. In spots where the bare basalt lava flows are exposed, note the striations carved into the surface by the last period of glaciation between 20,000 and 15,000 years ago.

In little over a mile the path abruptly steepens as it starts the final 550-foot ascent to the top of the Tumac Mountain cinder cone. After diagonaling up the final 300 feet of open slope, across a loose surface of tephra and red and black clinkers, the way arrives at the 6,300-foot-high summit. Below, at the base of the steep northwest slope is the 150-foot-deep crater, breached on its northwest rim, and in its heart lies a tiny blue lake. Trail #44 continues north, descending the volcano cone and shield to link with other trails to Twin Sisters Lakes, and on beyond to

Bumping Lake; however, the summit of Tumac Mountain is the destination of this trip.

The Tumac Mountain volcano is relatively young; it probably erupted between 30,000 and 20,000 years ago, spreading floods of olivine basalt down valleys on all sides of the vent within a 1½- to 5-mile radius. Lava flows accumulated to depths of over 700 feet, forming the base of today's Tumac plateau. Typical of basalt eruptions, the vent at Tumac Mountain erupted in a lava fountain that ejected incandescent drops of lava hundreds to thousands of feet in the air, where they coalesced and dropped to the ground as cinders that built the cone that stands above the crater.

Goose Egg Mountain

Road trip past a diorite intrusion

Elevation: Base of Goose Egg Mountain 2,880 feet, summit 4,566 feet
Maps: USGS Rimrock Lake, Tieton Basin; USFS Naches Ranger District
Driving Directions: 16.7 miles east of White Pass turn south on Tieton Road (FR 1200), and in 2 miles reach the southeast base of Goose Egg Mountain.

On the south side of Goose Egg Mountain is an impressive 1,000-foot-high rock wall sliced diagonally base-to-top by a dozen or more parallel

A maze of fracture lines spreads across the face of Goose Egg Mountain.

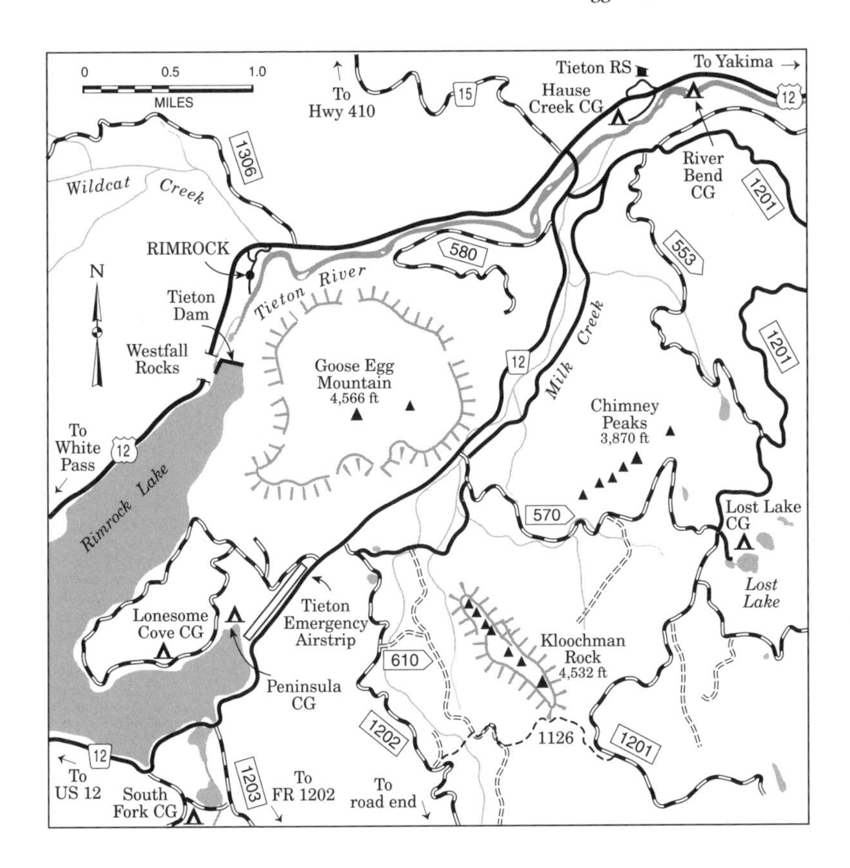

fracture lines. The rock between them has split into foot-thick slabs; the lower portions of the surface slabs have fallen away, while the upper parts overhang the wall beneath. At one spot along the base of the cliff, three 200- to 300-foot-high vertical rock flakes have separated from the main part of the face. Dedicated rock climbers would hike miles to reach such a challenge, yet this wall is scarcely 200 yards from a well-traveled forest road.

The mountain, which gets its name from the egg shape it presents from a distance, was not created from lava flows or volcanic sediments, as was much of the surrounding landscape. Instead, it is an intrusive body—a small diorite dome fed by a subterranean pool of magma—that pressed up through older layers of sedimentary rock sometime between 28 and 15 million years ago. The slabby surface on the south wall of Goose Egg Mountain was created when the heat of the intruding magma transformed the surrounding sedimentary rock into hornfels, a harder fine-grained rock. Millions of years of erosion by alpine glaciers, water, ice, and gravity have now exposed and shaped the mountain we see today.

Kloochman Rock and Chimney Peaks

Roads trips and hiking trails past andesite intrusions

Trail: Louie Trail #1126
Rating: (M); hikers, mountain bicycles, saddle and pack stock, motorcycles
Distance: 1.9 miles
Elevation: FR 1202 trailhead 3,440 feet, FR 1201 trailhead 4,170 feet
Maps: USGS Foundation Ridge, Tieton Basin; USFS Naches Ranger District
Driving Directions: 16.7 miles east of White Pass turn south on Tieton Road (FR 1200) (2-lane paved), and in 2.8 miles turn southeast on FR 1202 (1½-lane gravel). In 1.8 miles is the west end of Trail #1126. To reach the east end of the trail, at 1.8 miles from US 12, turn southeast from FR 1200 onto FR 1200570 (1½-lane gravel). In 2.5 miles turn south on FR 1201 (1½-lane gravel) and follow it for 1.8 miles to Trail #1126.

Kloochman Rock is a distinctive, ½-mile-long, sheer-faced stone slab that thrusts 1,200 feet above surrounding forested plateau like the dorsal fin of a giant dimetrodon. Northeast of Kloochman Rock are Chimney Peaks, a series of more subtle 50- to 100-foot-high pinnacles that stud a wooded ridge crest. These rocks, which lie about 2 miles south of US 12 near the east end of Rimrock Lake, are further examples of andesite that intruded into the base rock, probably from the same magma pool that fed the creation of Goose Egg Mountain and Westfall Rocks.

In his book *Of Men and Mountains,* the late U.S. Supreme Court Justice William O. Douglas devoted a chapter to Kloochman Rock, telling how his 1948 ascent of the rock by the relatively easy southwest route evoked memories of an earlier climbing experience on the rock in 1913. As an impetuous teenager, he and a friend, both with no climbing experience, set out to conquer the rock, first unsuccessfully via the southeast face, then successfully from the northwest. Both attempts nearly ended in disaster as the boys shed their clumsy shoes and climbed up narrow ledges in stocking feet, only to reach cul-de-sacs from which a hazardous retreat was but a hairbreadth from a fatal fall.

Douglas related how each managed to courageously save the other's life when ill-advised moves left each hopelessly stranded. His 1948 climb reawakened the excitement, fear, and terror of the earlier climb as well as the buoyant, boundless joy felt upon reaching the top. He wrote, "Kloochman was in my very heart. Here we had accomplished the impossible. We had survived terrible ordeals on her sheer walls. We had faced death down; and because of our encounter with it, we had come to value life more. On these dark walls in 1913 I had first communed with God."

Louie Trail #1126 links FR 1202 on the southwest side of Kloochman Rock to FR 1201 on its southeast side, in the process passing within 150 yards of the southeast end of the rock. The route leaves a pocket meadow trailhead off FR 1202 as a jeep road that continues north to a meadow with open views of the west side of Kloochman Rock. Although the views are great (and the USGS map indicates the trail goes this way), if you have

Kloochman Rock is a dramatic, 1,200-foot-high slab of andesite.

gone this far, you've missed the real trail. About 200 yards from the trailhead, a signpost obscured behind trees marks the spot where a narrow, poorly maintained section of trail heads east through trees along the south side of the previously mentioned meadow.

In 0.2 mile this path joins another abandoned road that switchbacks uphill to the east, with more viewpoints of the rock. At a Y-junction Trail #1126 leaves this old rutted roadbed and continues uphill as a rough hiker/biker path. In just under 0.5 mile the grade eases and breaks from trees, just a stroll away from the base of the south rump of Kloochman Rock. From here the path turns southeast into timber, crosses the head of a scree slope, and arrives at the FR 1201 trailhead.

The short chain of smokestack-shaped pinnacles that make up Chimney Peaks is only short cross-country hikes from FR 1200570. This road leaves FR 1200 1.7 miles south of US 12 then snakes up the drainage between Kloochman Rock and the Chimney Peaks ridge before joining FR 1201 just west of Lost Lake.

Kloochman Rock and Chimney Peaks are no easier for the inexperienced to climb today than they were when Justice Douglas made his first foray to Kloochman in 1913. Routes on these rocks and pinnacles are rated Class 3 to Class 4, which means that to ascend them safely climbing experience and, depending on the route, ropes and belaying are required. Much of the rock is loose and fractures easily, adding to the hazards of the climb. Do not attempt to climb them unless you have the skill and equipment to do so.

13

Mount Rainier

Mount Rainier, the highest point in the state of Washington, is the most outstanding example of a classic Cascades stratovolcano. The national park surrounding the mountain has over 80 miles of roads, some of which reach spectacular alpine viewpoints of the 14,411-foot peak at Paradise, on the south side of the mountain, and Sunrise, on its east side. The northwest corner of the 235,612-acre park is accessible via roads to Mowich Lake and the Carbon River. However, most of the road along the west side of the mountain has been closed by recurring debris torrents and probably will not be reopened to its previous terminus. Stevens Canyon Road links the east and west sides of the mountain. Along the east side of the park, Highway 123 joins Highway 410 at Cayuse Pass to US 12 east of White Pass. Short spur roads reach campgrounds at Ohanapecosh Hot Springs and White River.

The bulk of the park, however, belongs to walkers, hikers, and climbers. More than 300 miles of trails lie on the mountain, including the 93-mile-long Wonderland Trail that circles its base. Over twenty-five different climbing routes have been established to the summit, and each has several variations. The climbs range from the "relatively easy" (if ascending 9,000 vertical feet over volcanic debris and crevassed glaciers can be termed easy) to the "relatively insane" up near-vertical walls of ice-glued rubble, regularly raked by rockfall and avalanches. All recreational activities on Mount Rainier are well described in trail and climbing guides. Some of these guides are listed in Selected References, at the end of the book.

The park's roads and trails also offer outstanding opportunities to view various facets of the mountain's geological origins and history. The following description of the evolution of Rainier is offered to help visitors better appreciate the magnitude of the natural forces that have shaped the mountain and impacted surrounding areas and how these forces may continue to affect the mountain in the future.

Mount Rainier is a cone-shaped, composite volcano whose slopes are built up of many alternating layers of lava flows, tephra, ash, lahars, and volcanic debris that was deposited over hundreds of thousands of years of intermittent volcanic activity. The oldest formation in the base upon which the volcano was built is the Puget Sound group of sedimentary rocks, which was formed from eroded muds and sands that collected in the coastal region of the North American continent between 50 and 35 million years ago.

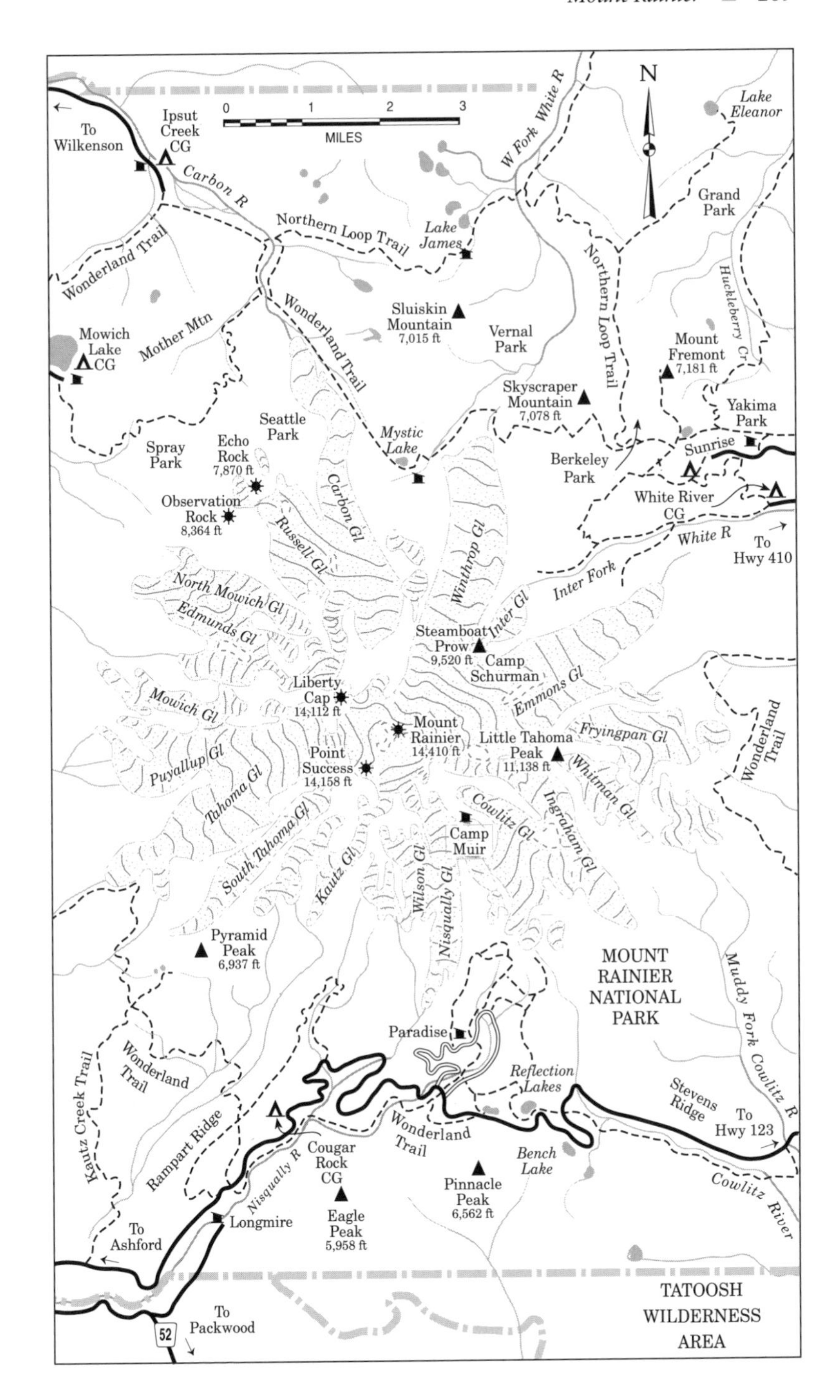
N
0 1 2 3
MILES
To Wilkenson
Ipsut Creek CG
Carbon R
Northern Loop Trail
Lake James
W Fork White R
Lake Eleanor
Grand Park
Wonderland Trail
Mowich Lake CG
Mother Mtn
Wonderland Trail
Sluiskin Mountain 7,015 ft
Vernal Park
Northern Loop Trail
Huckleberry Cr
Mount Fremont 7,181 ft
Skyscraper Mountain 7,078 ft
Yakima Park
Seattle Park
Mystic Lake
Spray Park
Echo Rock 7,870 ft
Berkeley Park
Sunrise
Observation Rock 8,364 ft
Carbon Gl
White River CG
Russell Gl
Winthrop Gl
White R
To Hwy 410
North Mowich Gl
Inter Fork
Edmunds Gl
Inter Gl
Steamboat Prow 9,520 ft
Camp Schurman
Emmons Gl
Liberty Cap 14,112 ft
Mowich Gl
Mount Rainier 14,410 ft
Little Tahoma Peak 11,138 ft
Fryingpan Gl
Wonderland Trail
Puyallup Gl
Point Success 14,158 ft
Whitman Gl
Tahoma Gl
Cowlitz Gl
Ingraham Gl
Camp Muir
South Tahoma Gl
Kautz Gl
Wilson Gl
Nisqually Gl
Pyramid Peak 6,937 ft
MOUNT RAINIER NATIONAL PARK
Muddy Fork Cowlitz R
Paradise
Reflection Lakes
Wonderland Trail
Kautz Creek Trail
Stevens Ridge
To Hwy 123
Rampart Ridge
Wonderland Trail
Cougar Rock CG
Bench Lake
Nisqually R
Pinnacle Peak 6,562 ft
Cowlitz River
Longmire
Eagle Peak 5,958 ft
To Ashford
TATOOSH WILDERNESS AREA
52
To Packwood

Near the end of this period, large volcanoes evolved on the newly accreted coastal plain of western Washington. Their eruptions produced lava flows and huge floods of volcanic rocks and mud that interbedded with the sands and silts brought by streams to the coast from the highlands of eastern Washington. Over the next 10 million years, multiple layers of andesite and dacite fragments, andesite lava flows, and other volcanic debris that were extruded built up to thicknesses of 7,500 to 10,000 feet. A newer and different surge of volcanic activity occurred some 30 to 25 million years ago; new flows of ash and volcanic rocks erupted and accumulated to depths of 450 to 3,000 feet, overlaying the consolidated volcanic sediments and erosion debris from the previous 8 million years. This activity occurred mostly to the south of the present volcano cone. About 5 million years later a third volcano began, and lava flows, mud flows, and erupted rocks were again layered over the existing base, this time to a depth of 1,000 to 3,000 feet. This volcano was located just east of the present park boundary.

A large bulb of magma rose beneath this 20-million-year-old layer cake of lavas and volcanic debris. About 25 million years ago fingers from this magma pool rose and squeezed up through cracks and fissures in the overlying rock. On the northeast side of Rainier the magma broke through the surface in a few spots as isolated lava flows; however, the bulk of this magma reservoir pressed upward, deformed the overlying rock formations, then congealed beneath their surface as a massive granodiorite and quartz monzonite pluton.

Volcanic activity in the vicinity apparently ceased for nearly 11 million years. During this time the pluton and its thick cap of old volcanic rock were lifted upward in gentle north–south-trending folds and were heavily eroded by channels cut through them by rivers and streams. This was the 2,300-foot-high base upon which the current mountain was built. The pluton indicates that Mount Rainier rose anew upon the roots of a larger and more ancient volcano.

The birth of Mount Rainier as we know it today began about a million years ago. Repeated eruptions poured andesite lava out of a central vent and down into deeply eroded channels on the vestigial mountain's flanks. These flows gradually filled the old stream channels with many thin layers of lava, stacked one on top of another, often interspersed with bands of tephra, lahars, and other volcanic debris. The streams and rivers that once ran down these drainages were forced to find paths around the edges of the channel-filling volcanic flows, inverting the former valleys into thick, high lava plateaus bounded by the newly cut watercourses.

A sequence of lava extrusions, tephra eruptions, and mud flows laced with lava fragments continued, slowly building the Rainier volcano cone to a height of 6,900 to 7,900 feet above the surrounding landscape. Periods of recurring glaciation also helped shape the embryonic giant. It is believed that by about 10,000 years ago the mountain reached a height of 15,500 feet, nearly 1,100 feet higher than the current summit. However, about 5,000 years ago an explosive ash eruption is thought to have triggered a tremendous avalanche, and the upper part of the cone slid away to the northeast. Liberty Cap, Point Success, and the top of Willis Wall are remnants of the rim of the resulting crater. A new volcano cone, Columbia Crest, largely filled the crater with lava and volcanic rubble, beginning about 2,500 years ago.

Most of the eruptions that built the mountain are thought to have been associated with lava fountains, in contrast to the violent nature of eruptions of nearby St. Helens. Periodically, however, Rainier would clear its throat and belch out pumice and tephra. One of the older of this type of eruption, which occurred between 70,000 and 30,000 years ago, dumped a thick pumice layer over the landscape to the east, northeast, and southeast. About this same time, two smaller satellite volcanoes erupted at the northwest base of the mountain; their vents are marked by today's Echo and Observation rocks. However, each of these is unique; Echo Rock is composed primarily of cinders that built up a cone around the lava fountain vent, while Observation Rock is mostly solidified magma, with its cinder cover eroded away.

In the last 10,000 years there have been at least eleven explosions of ash and tephra from the mountain, one as recent as between 1820 and 1854. Most of these eruptions were fueled by steam generated at the contact of hot magma with ground water from melting glaciers. The largest of these explosions, which ejected 392 million cubic yards of pumice and volcanic fragments, occurred about 2,500 years ago. This event accompanied the volcanic activity that built Columbia Crest, the most recent of the three craters atop the mountain.

Intermittent periods of heavy continental and alpine glaciation encompassed Mount Rainier (and much of Washington) during the past 150,000 years. The already none-too-solid structure of the mountain's rock was modified and weakened by the water stored in its glaciers, heat found in its magma core, and chemical changes wrought by the combination of the two. At least fifty-five lahars and landslides have been noted occurring on the mountain in the past 7,000 years. Several of these slides are thought to correlate with the tephra explosions mentioned earlier, including the largest, the Osceola debris flow, which occurred about 5,000 years ago. This slide originated high on the northeast side of the mountain, above the present summit, in the vicinity of the previous summit, whose collapse spawned the flow. It swept down the flank of the mountain into the White River and West Fork of the White River drainages and continued downstream for more than 45 miles, to the present-day site of Auburn, covering the White River valley to depths of between 30 and 500 feet. This flow alone spread over more than 62 square miles of Puget Sound lowlands, and it, and its subsequent erosion, caused the Puget Sound shoreline to shift seaward from 17 to 31 miles, filling the Duwamish valley, which had previously been an arm of the sound. The volume of the Osceola flow was roughly 60 times that of the disastrous mud flow that swept down the

Glaciers on Mount Rainier and its satellite peak, Little Tahoma, shown here, continue to shape the mountain.

North Fork of the Toutle to the Columbia River during the 1980 Mount St. Helens eruption. Hydrothermal activity in the old crater has created a northwest-tilting clay layer in the bowl below beneath the present summit, Columbia Crest, which is thus poised to slide northeast at some future date. Think of that when you build your dream house in the White River flood plain!

Most of the changes to Mount Rainier in recent history have been floods and mud flows caused by glacial melt. Major floods from the Nisqually Glacier occurred in 1926, 1932, 1934, and 1955; a large slurry flood surged down the Kautz River in 1947; and in 1963 a massive debris avalanche peeled from the north side of Little Tahoma and raced down the Emmons Glacier to White River Campground. Repeated hot summers in the 1980s especially impacted the southwest-facing South Tahoma Glacier, which has become stagnant. The glacier somehow manages to entrap a large volume of water, which periodically breaks loose and sweeps down Tahoma Creek carrying with it huge volumes of rock, sand, and silt. At least twenty-one such debris flows have moved down Tahoma Creek since 1962, destroying a campground, erasing the lower end of the Tahoma Creek trail, and blocking West Side Road.

Although Rainier's history of eruptions has been relatively non-violent, the mountain is still geologically active. Over twenty significant earthquakes are recorded every year, more than any other composite Cascade volcano other than St. Helens. The famous steam caves in the summit crater attest to a hot magma source not too far below the surface, and there are at least seven other known thermal areas above 11,000 feet. Its history of mud flows and debris avalanches has certainly had a significant impact on the surrounding geography, and continued deterioration of the mountain by glaciation, snow melt, and magma-heated steam can only portend more of the same. Whether a major landslide could trigger a massive explosion of its magma core, such as occurred on St. Helens in 1980, is subject to speculation. At any rate, it's a safe bet that the Pacific Northwest has not heard the last from this somnolent giant.

Grand Park

Hiking trail to an early canyon-filling lava flow

Trails: Wonderland Trail, Northern Loop Trail, Grand Park Trail
Rating: (M); hikers
Distance: 7.2 miles
Elevation: Trailhead 6,350 feet, Grand Park 5,696 feet
Driving Directions: Take Yakima Park Road southwest from Highway 410, 4.5 miles south of the park boundary. At the White River Campground junction in 5.2 miles, continue uphill on Yakima Park Road another 10.1 miles to Sunrise.

Mount Rainier's "parks" are broad subalpine meadows dotted with sparse stands of stunted mountain hemlock and subalpine fir that manage to survive in the harsh climate and short growing season found in their mile-high environment. But when the snow melts these parklands blaze with color from a bonanza of wildflowers that flourish here: avalanche lilies, monkeyflower, beargrass, shooting star, and columbine are but a few of the seventy-six species that have been identified. Three of these stunning parks, Yakima, Berkeley, and Grand, are traversed in this hike.

The broad flat bench of Grand Park is a striking example of inverted topography. When vast outpourings of andesite lava flows began forming present-day Rainier about a million years ago, this area had already been eroded by the deep channel of the ancient Grand Park River. The new lava flows, which were funneled down this river canyon, filled it layer by layer. The displaced river eroded new channels (today's Huckleberry Creek and the West Fork of the White River) along the edges of this massive flow, causing the one-time canyon to become a broad, high ridge.

To reach Grand Park from Sunrise, take the Wonderland Trail west past Frozen Lake to Berkeley Park, where it joins the Northern Loop Trail. En

The broad plateau of Grand Park is an ancient valley that was filled with lava flowing from the volcanic vent of Mount Rainier.

route you pass over another canyon-filling lava flow, which runs from Burrows Mountain through Yakima Park to Sunrise Point. As you head north through the flower meadows of Berkeley Park, the terrain crossed is an old glacial moraine. Skyscraper Mountain, to the west, shows many horizontal outcrops of sills, intruded sheets of lava from a dike and sill complex that capped the underlying Tatoosh pluton.

From the saddle at the north end of Berkeley Park, the trail climbs gradually to the broad open bench of Grand Park, the top of the old canyon-filling lava flow. In good weather there is a fantastic view north to the confluence of the Emmons and Winthrop glaciers, the source of the huge Osceola mud flow. Imagine in your mind's eye that the mountain you are looking at is 1,500 feet higher and, after an eruption at the summit, the entire glacier-clad mountainside above suddenly gives way and slides downhill, splitting around the ridge on which you are standing, and surging downstream to Puget Sound. It happened. And it could happen again!

Mount Rainier Summit Crater

Climb to summit fumeroles

Trails: Ingraham Glacier–Disappointment Cleaver route: Skyline Trail and Pebble Creek Trail from Paradise to Pebble Creek; Emmons Glacier route: Glacier Basin Trail from White River Campground to Glacier Basin

Rating: Skyline and Pebble Creek trails (E), Glacier Basin Trail (M), snowfield and glacier travel above Panorama Point and Glacier Basin (D); hikers

Distance: Ingraham Glacier–Disappointment Cleaver route, Paradise to summit 8.6 miles; Emmons Glacier Route, White River Campground to summit 10.2 miles

Elevation: Paradise 5,380 feet, White River Campground 4,310 feet, summit 14,411 feet

Maps: USGS Mount Rainier East, Mount Rainier West, Sunrise; NPS Mount Rainier

Driving Directions: To reach Paradise, take Highway 706 east from Elbe for 13 miles to the Nisqually Entrance to Mount Rainier National Park, then continue east past Longmire to Paradise, 18.5 miles. To reach White River Campground, take the White River Road west from Highway 410, 3.5 miles north of Cayuse Pass, and follow it 7 miles to the campground.

Note: *Travel anywhere on glaciers or above Camp Muir requires registration with park rangers as a climber and checkout upon return. Both climbing routes mentioned here involve glacier travel and require mountaineering skills and experience.*

For the penultimate in volcanic craters in the South Cascades, it's difficult to beat the summit of 14,411-foot Mount Rainier. Here you get not just a crater but also a view from the highest point in the state and a close-up view of the steam caves that prove there's still fire a'breathing in the core of the mountain. All this is not achieved by a Sunday stroll. The climb of the mountain is a serious and very strenuous effort, because you gain more than 1¾ miles of elevation and more than half the distance is across steep, crevassed glaciers. About half the 7,000 to 8,000 people who attempt the climb annually fail to reach the summit, due to altitude problems, bad weather, or lack of physical conditioning.

The two most popular routes to the summit are from the south side of the mountain from Paradise via the Ingraham Glacier and from the northeast side of the mountain from White River Campground via the Inter and Emmons glaciers. Even these routes are challenging and have recorded fatalities. Both require glacier climbing equipment (crampons, rope, ice ax, sun goggles, warm clothing), prior experience with glacier travel, and good physical conditioning to climb safely. The concessionaire guide

The crater of Columbia Crest on the summit of Mount Rainier; Liberty Cap is on the right

service requires a one-day class on glacier travel and safety prior to participating in any guide-led climb. Fred Beckey's book *Cascade Alpine Guide, Vol. 1: Columbia River to Stevens Pass* (Seattle: The Mountaineers, 1989) provides complete climbing details for all routes.

Either of the two routes mentioned here take you to Columbia Crest, the highest point on the mountain. Its crater contains one of the three fumaroles on the rim of the broad summit cap. One of the other fumaroles (at Liberty Cap) lies ½ mile to the northwest across the broad summit; near Point Success, ¼ mile southwest, is a third fumarole. All three fumaroles lie on the rim of what was once the primary caldera of a younger, 1,500-foot-higher volcano. During summer months the northeast lip of Columbia Crest melts free of snow. Steam created from the contact of glacial ice with the hot inner core of the volcano melts out caves between the crater-filling ice and the pumice rim. These steam caves have saved the life of more than one climbing party trapped on the summit by bad weather.

Glossary

andesite: A medium-dark, fine-grained, extrusive, igneous rock. Andesite is roughly 60 percent silica and is predominately composed of feldspar, with some hornblende and biotite. The lava originates from the melting of subducting oceanic crust and is already crystallized in the magma at the time of eruption.

ash: Very fine-grained dust or volcanic particles explosively erupted from a volcano vent.

basalt: A dark, fine-grained, extrusive, igneous rock. Basalt is roughly 50 percent silica and is rich in ferromagnetic silicates and feldspar. The lava is created from the partial melting of the earth's upper mantle. Molten basalt flows easily and erupts non-violently.

cinder cone: A usually small volcano cone composed of solid fragments ejected from a central vent by a lava fountain. Fragments range from small foamy andesite or basalt cinders to watermelon-sized volcanic bombs. Tongues of basalt are often extruded from the base of the same vent.

cirque: A glacially carved, steep-walled bowl at the head of a mountain valley.

clinker: A small fragment of lava ejected from a lava fountain that congeals to solid rock while still in mid-air.

columnar basalt: A basalt lava flow that, because of rapid cooling, shrinks and splits into columns, typically hexagonal in shape.

composite volcano (stratovolcano): A high, steep-sided, cone-shaped volcano built up over an extended period from alternating layers of lavas, ash flows, and explosive eruption debris.

dacite: A light-colored, fine-grained, extrusive, igneous rock. Dacite is roughly 65 percent silica, and it often includes light and dark scattered crystals. The lava originates from a mix of upper mantle magma and reprocessed crust. Molten dacite is slow flowing and often erupts explosively.

debris avalanche: A high-velocity, downslope movement of loose, unsorted rock debris.

diabase: A dark, coarse-grained, intrusive, igneous rock of basaltic composition. Its feldspar crystals grow first, with the space between them later filled by darker minerals. Diabase cools very slowly at considerable depth beneath the surface, allowing time for large crystals to grow.

dike: A relatively thin, homogeneous, igneous rock formed when magma is injected across bedding planes through a crack in overlying rock bodies.

diorite: A black-and-white speckled, coarse-grained, intrusive, igneous rock. Diorite is roughly 60 percent silica, and it is composed of feldspar and darker minerals such as hornblende and biotite. The rock is similar in appearance to granite, but darker.

dome: A steep-sided, bulb-shaped mass of extremely viscous lava such as dacite or rhyolite. The surface often fractures into blocks that form a rubble cone around the vent plug.

erosion: The process by which rocks are loosened or dissolved and removed from their original location. Sources of erosion include weathering, glaciation, earth movements, running water, wave-caused hydraulics, and gravity.

fissure: An extensive fracture in rocks.

fumarole: A volcano vent emitting steam and gasses.

gabbro: A very dark, coarse-grained, intrusive, igneous rock. Gabbro is roughly 48 percent silica and is composed of calcium-bearing feldspar and ferromagnetic silicates. Gabbro cools slowly at great depth below the surface, allowing the growth of large crystals.

granite: A light varicolored, coarse-grained, intrusive, igneous rock. Granite is more than 60 percent silica and is composed of quartz, feldspar, biotite, and hornblende. It is the chief intrusive igneous rock in the continental crust. This is the magma equivalent of rhyolite.

granodiorite: A coarse-grained, intrusive, igneous rock similar to granite. Granodiorite is roughly 65 percent silica with at least a 1:5 quartz–feldspar ratio. This is the magma equivalent of dacite.

hornblende: A common silicate mineral that contains combinations of calcium, sodium, potassium, magnesium, aluminum, and iron.

intrusive rock: An igneous rock produced when magma solidifies and hardens underground. Intrusive rocks are generally coarse-grained and produce such features as plutons, stocks, dikes, and sills.

lahar (mud flow): A rapidly moving mixture of pyroclastic debris, ash, and water that originate from a volcano.

lava: Molten rock extruded or ejected from volcano vents.

lava fountain: Lava ejected into the air from a volcano vent. It is visually analogous to a water fountain.

lava plug: The harder lava of the central vent system of a volcano that remains after softer outer slopes have eroded away.

magma: Subsurface molten rock derived from parts of the earth's upper mantle or lower crust or from rock melted in subduction zones. Magma chambers feed volcanoes through dikes, rifts, and sills.

mantle: The major portion of the earth's interior, about 1,800 miles thick, that lies between the planet's core and its crustal rock.

monzonite: A coarse-grained, intrusive, igneous rock similar to granite, but containing less quartz.

mud flow (*see* lahar)

olivine: A group of silicate minerals of varying compositions that are rich in iron and magnesium. It is a constituent of the earth's mantle.

pahoehoe: A ropy, wrinkled form of basalt formed by molten, fast-flowing lava.

parasitic volcano: A volcano vent on the apron of a major composite volcano that derives its magma from the same pool as the volcano.

pillow basalt: A coarse, bulbous basalt, resembling pillows in shape, that is formed by eruptions of lava under water.

pluton: A large mass of intrusive, igneous rock, formed from magma rising from the mantle, that cools and crystallizes before reaching the surface of overlying rock. It often represents the unroofed magma chamber of an ancient volcano.

pumice: A frothy lava filled with gas bubbles that expand upon eruption to form a rock so light that it can float on water.

pyroclastic eruption: A violent, explosive volcanic eruption that ejects a mass of hot debris and gasses that move laterally at high velocity across the ground surface or solidify in mid-air to rock fragments.

rhyolite: A pale, eruptive, igneous rock of very high viscosity. Rhyolite is roughly 70 percent silica, has the consistency of dry peanut butter, and tends to erupt explosively, generating pumice, ash, and larger volcanic fragments.

scarp: A line of cliffs marking the upthrown side of a fault (a vertical break in the earth's crust).

seamount: A submarine volcano rising above the ocean floor.

sedimentary rock: Rocks formed by the compacting or cementing of muds, sands, plant and animal remains, and dissolved minerals.

shield volcano: A broad, low-relief volcano with a gently sloping dome, generally formed from highly fluid basalt that flows far from the source vent before cooling.

silica: A combination of silicon and oxygen (SiO_2) that is a chemical component of various common minerals.

sill: A relatively thin sheet of intrusive rock that forms when magma is injected between bedding layers of the intruded rock.

slurry: A watery mixture of insoluble matter, such as mud.

strata: Horizontal sheets of rock separated from each other by bedding planes.

stratovolcano (*see* composite volcano)

subduction: The process by which the edge of an oceanic plate is drawn down beneath a lighter plate of continental crust to be reconsumed by the underlying mantle. When the subducted plate reaches a depth of about 60 miles, it releases water, which then interacts with the overlying mantle to produce basalt. The basalt rises and is trapped beneath the overlying crust to form magma chambers.

talus: An accumulation of rock debris eroded from and found at the base of cliffs.

tectonic plate: A slab of oceanic crust, continental crust, or both that is coupled to a relatively rigid slab of the upper mantle of the earth. The plate "floats" on the underlying fluid layers of the mantle.

tephra: A general term covering rocks and mid-air solidified lava particles ejected during a volcanic eruption.

tuff: A fine-grained rock composed of compacted or cemented volcanic ash. Tuff is often classified as a sedimentary rock.

tumulus: A hollow egg-shaped bulge on the surface of lava tube fields caused by pressure from gas released from the lava. It most often collapses to form a small, shallow crater.

tuya: A volcano that initially erupted beneath either glacial ice or meltwater lakes to form a meltwater lake, built a cone through it, and then added to its height with further lava flows above the glacier meltwater surface.

Appendix

Addresses and Telephone Numbers

Washington State Department of Natural Resources, Southwest Region, 601 Bond Road, Box 280, Castle Rock, WA 98611-0280. 1-800-527-3305

Gifford Pinchot National Forest

Supervisor's Office, 6926 East Fourth Plain Boulevard, Vancouver, WA 89669-8944. (360) 696-7500

Randle Ranger District, 10024 Highway 12, Randle, WA 98377. (360) 497-7565

Packwood Ranger District, Highway 12, Packwood, WA 98361. (360) 495-5515

Mount Adams Ranger District, 2455 Highway 141, Trout Lake, WA 98650. (509) 395-2501

Wenatchee National Forest

Supervisor's Office, 301 Yakima Street, P.O. Box 811, Wenatchee, WA 98807. (509) 662-4335

Naches Ranger District, 10061 Highway 12, Naches, WA 98937. (509) 653-2205

National Parks, Monuments, and Scenic Areas

Mount St. Helens National Volcanic Monument, Route 1, Box 369, Amboy, WA 98601. (360) 247-5473

Columbia Gorge National Scenic Area, Suite 200, 902 Wasco Avenue, Hood River, OR 97031. (503) 386-2333

Mount Rainier National Park, Tahoma Woods, Star Route, Ashford, WA 98304. (360) 569-2211

Selected References

Beckey, Fred. *Cascade Alpine Guide, Vol. 1: Columbia River to Stevens Pass,* 2nd edition: Seattle, Washington, The Mountaineers, 1989.

Mueller, Marge and Ted. *Exploring Washington's Wild Areas:* Seattle, Washington, The Mountaineers, 1994.

Pringle, Patrick T. *Roadside geology of Mount St. Helens National Volcanic Monument and vicinity:* Washington State Department of Natural Resources, 1993.

Schaffer, Jeffrey P., and Selters, Andy. *The Pacific Crest Trail, Vol. 2: Oregon and Washington:* Berkeley, California, Wilderness Press, 1992.

Seesholtz, David. *Mount St. Helens, Pathways to Discovery:* Vancouver, Washington, A Plus Images, 1993.

Spring, Ira, and Manning, Harvey. *50 Hikes in Mount Rainier National Park:* Seattle, Washington, The Mountaineers, 1988.

———. *100 Hikes in Washington's South Cascades and Olympics,* 2nd edition: Seattle, Washington, The Mountaineers, 1992.

Index

About the Authors

MARGE and TED MUELLER are outdoor enthusiasts and environmentalists who have explored Washington state's waterways, mountains, forests, and deserts for nearly forty years. Ted has taught classes on cruising in Northwest waters, and both Marge and Ted have instructed mountain climbing through the University of Washington. They are members of The Mountaineers and The Nature Conservancy, and Ted is a board member of the Washington Water Trails Association.

For this book they teamed up with photographer IRA SPRING, who has introduced legions of future environmentalists to the Northwest wilderness with the *100 Hikes in*™ guidebooks, as well as numerous other volumes.

THE MOUNTAINEERS, founded in 1906, is a nonprofit outdoor activity and conservation club, whose mission is "to explore, study, preserve, and enjoy the natural beauty of the outdoors. . . ." Based in Seattle, Washington, the club is now the third-largest such organization in the United States, with 15,000 members and four branches throughout Washington state.

The Mountaineers sponsors both classes and year-round outdoor activities in the Pacific Northwest, which include hiking, mountain climbing, ski-touring, snowshoeing, bicycling, camping, kayaking and canoeing, nature study, sailing, and adventure travel. The club's conservation division supports environmental causes through educational activities, sponsoring legislation, and presenting informational programs. All club activities are led by skilled, experienced volunteers, who are dedicated to promoting safe and responsible enjoyment and preservation of the outdoors.

The Mountaineers Books, an active, nonprofit publishing program of the club, produces guidebooks, instructional texts, historical works, natural history guides, and works on environmental conservation. All books produced by The Mountaineers are aimed at fulfilling the club's mission.

If you would like to participate in these organized outdoor activities or the club's programs, consider a membership in The Mountaineers. For information and an application, write or call The Mountaineers, Club Headquarters, 300 Third Avenue West, Seattle, Washington 98119; (206) 284-6310.

Send or call for our catalog of more than 300 outdoor titles:

The Mountaineers Books
1001 SW Klickitat Way, Suite 201
Seattle, WA 98134
1-800-553-4453